T0211962

Electromagnetic Scattering Using the Iterative Multiregion Technique

© Springer Nature Switzerland AG 2022
Reprint of original edition © Morgan & Claypool 2007

All rights reserved. No part of this publication may be reproduced, stored in a retrieval system, or transmitted in any form or by any means—electronic, mechanical, photocopy, recording, or any other except for brief quotations in printed reviews, without the prior permission of the publisher.

Electromagnetic Scattering Using the Iterative Multiregion Technique
Mohamed H. Al Sharkawy, Veysel Demir, and Atef Z. Elsherbeni

ISBN: 978-3-031-00574-9   paperback
ISBN: 978-3-031-01702-5   ebook

DOI: 10.1007/978-3-031-01702-5

A Publication in the Springer series
SYNTHESIS LECTURES ON COMPUTATIONAL ELECTROMAGNETICS #19

Lecture #19

Series Editor: Constantine A. Balanis, Arizona State University

Series ISSN
ISSN 1932-1252   print
ISSN 1932-1716   electronic

Electromagnetic Scattering Using the Iterative Multiregion Technique

Mohamed H. Al Sharkawy
Arab Academy for Science and Technology and Maritime Transport
Alexandria, Egypt

Veysel Demir
Northern Illinois University

Atef Z. Elsherbeni
The University of Mississippi

SYNTHESIS LECTURES ON COMPUTATIONAL ELECTROMAGNETICS #19

ABSTRACT

In this work, an iterative approach using the finite difference frequency domain method is presented to solve the problem of scattering from large-scale electromagnetic structures. The idea of the proposed iterative approach is to divide one computational domain into smaller subregions and solve each subregion separately. Then the subregion solutions are combined iteratively to obtain a solution for the complete domain. As a result, a considerable reduction in the computation time and memory is achieved. This procedure is referred to as the iterative multiregion (IMR) technique.

Different enhancement procedures are investigated and introduced toward the construction of this technique. These procedures are the following: 1) a hybrid technique combining the IMR technique and a method of moment technique is found to be efficient in producing accurate results with a remarkable computer memory saving; 2) the IMR technique is implemented on a parallel platform that led to a tremendous computational time saving; 3) together, the multigrid technique and the incomplete lower and upper preconditioner are used with the IMR technique to speed up the convergence rate of the final solution, which reduces the total computational time. Thus, the proposed iterative technique in conjunction with the enhancement procedures introduces a novel approach to solve large open-boundary electromagnetic problems including unconnected objects in an efficient and robust way.

KEYWORDS

finite difference frequency domain, electromagnetic scattering, iterative multiregion technique, multigrid technique, preconditioner

Contents

CHAPTER 1

Introduction

Numerical analyses of large-scale electromagnetic problems require long computational time and large computer memory. One of the goals of ongoing computational electromagnetic research is to develop time and memory-efficient algorithms. A class of time and memory-efficient algorithms divides the computational domain into smaller subdomains and then combines the subdomain solutions after introducing the effect of interactions between these subdomains. A group of methods that decomposes the computational domain into subdomains is known as the domain decomposition method (DDM) [1–17]. These methods in general require common boundaries between subdomains and boundary conditions are enforced on subdomain interfaces. There are usually two approaches used with the applications of the coupling effects: the direct method imposes the continuity of the fields on the partition interfaces and generates a global coupling matrix [17], whereas the iterative method [1, 4] ensures the coupling between the adjacent elements by the transmission condition (TC) as described in [1]. It is possible to solve each subdomain with the same method such as with finite element method [4] or finite difference frequency domain (FDFD) method [8]. However, some domain decomposition methods have the flexibility that, in each subdomain, the most efficient method can be used independently to solve Maxwell's equations [7]. Therefore, the complexity of the problem can be reduced, and a time and memory efficiency algorithm can be achieved. Another advantage of the domain decomposition methods is that they are suitable to develop parallel processing techniques [14–16] and thus enable highly scalable algorithms.

To economically provide efficient solution to large-scale electromagnetic problems, especially those that involve open boundaries such as the scattering from multiple objects, decomposing the computation domain into separate subregions would be preferable. It is then necessary to develop accurate procedures to support the interaction between the unconnected subregions. Some hybrid techniques based on combinations of method of moments (MoM), finite element, finite difference time domain (FDTD) and physical optics have been used to solve a class of these problems, in which part of the problem is usually large compared with other parts [18–20].

In this book, a new technique based on the FDFD method and an iterative procedure to calculate the scattering from multiple objects similar to that described in [21] are presented. In this approach, the problem is decomposed into separated subregions, each subregion containing a scatterer or a group of scatterers. The scattered electromagnetic near fields are calculated because of the incidence of a time-harmonic wave in each subregion, using the FDFD method or any other appropriate method. Then, fictitious electric and magnetic currents on imaginary surfaces surrounding the objects in these subregions are calculated, using the equivalence principle. Radiated fields by these currents are then considered as incident fields on the opposing subregions. The same procedure of calculating the subregion field components, the fictitious currents, and the radiated fields on the opposing domains is repeated iteratively until a convergence is achieved.

The procedure presented in this work, referred to as iterative multiregion (IMR) technique, requires a solution of fields in the subregions a number of times instead of one solution of the complete domain. This technique effectively reduces the size of the required memory. Furthermore, the CPU time reduction can be achieved if the separation between some subregions is large and/or coarser grids are used in some of the subregions, which may not be possible to use if only one domain is used for the solution of the original complete problem. Another unique feature of the IMR technique is that it can provide solutions to large-scale problems that are difficult or impossible to solve using direct methods. The application of this technique is performed on two- and three-dimensional (2D and 3D, respectively) scatterers, and the verification of the generated numerical data from the FDFD code is performed. The use of the FDFD provides the flexibility in defining composite (in shape and type) structures in each subregion. It also provides a much stable solution relative to other available methods and a more convenient procedure for performing the interaction between the subregions based on well-known theorems. The use of the FDTD technique was intentionally avoided in this work as the interaction process between subregions would involve complex bookkeeping for the source and its effect on the spatial and time domains from one subregion to the other. Despite that both FDFD and FDTD methods share a discretization constraint, the latter requires more attention regarding the choice of the time step and the parameters of the source time domain waveform. Furthermore, the FDFD is free of dispersion, which is one of the drawbacks of the FDTD technique. The FDFD solution provided in this work has some similarities to the 2D FDFD analysis in [22]. For the 3D analysis, the present formulation only uses the three field components, E_x, E_y, and E_z, instead of using the traditional six field components, E_x, E_y, E_z, H_x, H_y, and H_z, as one would expect following the formulation presented in [22]. Thus, as will be shown later, this FDFD formulation allows for additional memory saving.

The FDFD equations are constructed based on the scattered field (SF) formulation, that is, the total field (TF) is assumed to be the sum of incident, and SFs. The use of these equations requires the computation of incident field components at all grid nodes in the computation space. However, it has been realized that a considerable amount of computation time is spent for the calculation of the incident field components in a region due to the fictitious current sources in the other regions. Therefore, two ideas were proposed in this work to speed up this calculation: an interpolation process that can be simply described as an averaging process and the idea of using the TF/SF formulation [23].

A hybrid technique is also presented in this work, which combines the desirable features of two different numerical methods, FDFD and the MoM, to analyze large-scale electromagnetic problems. This is done by using each appropriate technique for a subregion and then applying an iterative procedure between the two solutions to calculate the scattering from multiple objects that are defined in the designated subregions. The idea behind applying this hybrid technique is to show the flexibility of the IMR technique in combining different solutions for each subregion to reach the desired solution in the most efficient and economical way.

Further investigations were done in this book to introduce some enhancing procedures to the IMR solution, thus providing a more robust technique used to solve large electromagnetic problems, within reasonable time and with the least memory usage. The idea of introducing multiple processors to the IMR solution is also proposed and developed, where the subregions are solved simultaneously, each on a separate processor. Thus, a remarkable CPU time saving can be achieved. The IMR technique is basically proposed to solve large electromagnetic problems in an efficient way; these problems usually generate a large number of unknowns that are represented in one matrix form. A direct matrix inversion solution would normally be very difficult if not impossible to achieve on current generation of moderate computer systems. Hence, iterative solvers are considered as an alternative to provide a solution for these set of linear equations. The multigrid technique, as an initial guess for the solution, and the incomplete lower and upper (ILU) triangular matrix factorization, as a preconditioning, are thus considered as a robust way to provide an approximate solution to the iterative solver. This results in accelerating the rate of convergence of the used iterative solvers for the resulting FDFD matrix equation and thus speeding up the solution process. More explanation on these points will be provided in separate chapters of this book.

• • • •

CHAPTER 2

Basics of the FDFD Method

This chapter presents a brief introduction to the FDFD method, which is an extended version to the finite difference method. The construction of the finite difference method from a given differential equation will thus be illustrated, which essentially involves estimating derivatives numerically. Furthermore, since the first step in constructing an FDFD algorithm is to discretize the computational domain into cells and define the locations of the electric and magnetic field vectors on each cell; the Yee cell will also be presented in this chapter where the locations of the field vector components are defined. A perfectly matched layer (PML), as an absorbing boundary, is used to terminate the computational domain.

2.1   FDFD METHOD

Most of the time, it is hard to find an analytical solution to real world electromagnetic problems. Classical analytical approaches may fail if the partial differential equation (PDE) is not linear and cannot be linearized without seriously affecting the result, the solution region is complex, the boundary conditions are of mixed types, or the medium is inhomogeneous or anisotropic.

Whenever a problem of such complexity arises, numerical solutions must be considered. The finite difference method was one of the first techniques used to solve differential equations numerically due to the advantage of being easily understood and implemented. Nowadays, the techniques known as FDTD and FDFD methods are the most popular finite difference formulations for solving electromagnetic problems in both time and frequency domains. Although FDFD is theoretically very simple and easy to program, the method is considered quite efficient to provide a much more stable solution relative to other methods. The FDFD method has been widely used in the solutions of various problems. Numerous publications are based on this method. This book is also based on the FDFD method, where it has been used to solve Maxwell's equations in the frequency domain. In this chapter, the basics of the method are introduced so that the reader will be better equipped to comprehend the formulations presented in subsequent chapters.

The FDFD method is based on approximating the spatial derivative by finite differences. These finite difference approximations are algebraic in form; they relate the value of the dependent variable at a point in the solution region to the values at some neighboring points. Given a function

$f(x)$ shown in Figure 2.1, its derivative at point x_0 can be approximated based on three different schemes:

$$\frac{\partial f(x_0)}{\partial x} = f'(x) \approx \frac{f(x_0 + \Delta x) - f(x_0)}{\Delta x}, \tag{2.1}$$

$$\frac{\partial f(x_0)}{\partial x} = f'(x) \approx \frac{f(x_0) - f(x_0 - \Delta x)}{\Delta x}, \tag{2.2}$$

$$\frac{\partial f(x_0)}{\partial x} = f'(x) \approx \frac{f(x_0 + \Delta x) - f(x_0 - \Delta x)}{2\Delta x}. \tag{2.3}$$

Equation (2.1) is the forward difference scheme, (2.2) is the backward scheme, and (2.3) is the central difference scheme.

Considering the Taylor's series expansions of $f(x + \Delta x)$ and $f(x\ \Delta x)$

$$f(x + \Delta x) = f(x) + \Delta x f'(x) + \frac{(\Delta x)^2}{2!} f'' + \frac{(\Delta x)^3}{3!} f''' + \ldots \tag{2.4}$$

$$f(x + \Delta x) = f(x) + \Delta x f'(x) + \frac{(\Delta x)^2}{2!} f'' + \frac{(\Delta x)^3}{3!} f''' + \ldots \tag{2.5}$$

and by taking the difference of (2.4) and (2.5) and dividing by $2\Delta x$, one obtains

$$f'(x) = \frac{f(x + \Delta x) - f(x - \Delta x)}{2\Delta x} - \frac{(\Delta x)^2}{6} f'''(x) + \ldots \tag{2.6}$$

The first term on the right side of (2.6) is the central difference approximation to $f'(x)$ as given in (2.3), and the other terms are the errors introduced by truncating the series. This error is

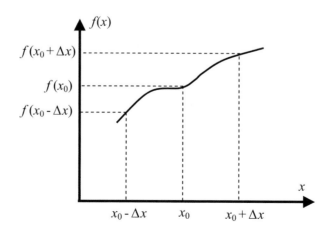

FIGURE 2.1: Estimates for the derivative of $f(x)$ at discrete points using forward, backward, and central differences.

of the order of $(\Delta x)^2$ or simply $O(\Delta x)^2$. The error is proportional to the square of the finite difference Δx; therefore, the central difference scheme is considered second-order accurate. In the same manner, it can be shown that the forward difference and backward difference schemes are first-order accurate. Although it is possible to obtain and use more accurate schemes, the second-order accurate central difference scheme is accurate enough to use in most of practical electromagnetic applications. The details of the method can be found and studied from many other sources [24, 25].

2.2.   THE YEE CELL

The first step in constructing an FDFD algorithm is to discretize the computational domain into a number of cells. Electric and magnetic field components are associated with these cells; thus, the definition of the location of these field components on each cell has to be considered. Yee [26] constructed an algorithm that solves for both electric and magnetic fields using Maxwell's curl equations, where he was able to define the location of the fields in a staggered fashion as shown in Figure 2.2.

Based on Yee's definition, there are three electric and magnetic field components oriented in such a way that every electric field component is surrounded by four circulating magnetic field components, which depicts Ampere's law. Similarly, every magnetic field component is surrounded by four circulating electric field components, which depicts Faraday's law. Based on the Yee space lattice, one can fit these field components to the FDFD expressions when the second-order accurate central difference scheme is used to discretize the space derivatives in Maxwell's curl equations.

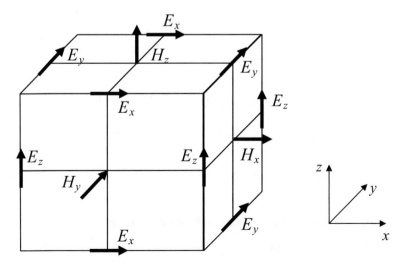

FIGURE 2.2: Positions of the electric and magnetic field vector components on the Yee space lattice.

2.3   3D FDFD FORMULATION

Starting from Maxwell's equations for the total electric and magnetic fields for time harmonic convention $e^{j\omega t}$

$$\nabla \times \overline{E}^{tot} = -j\omega\mu\,\overline{H}^{tot}, \quad \nabla \times \overline{H}^{tot} = +j\omega\varepsilon\,\overline{E}^{tot}. \tag{2.7}$$

Then, by separating the TF into incident and SF components, we obtain

$$\nabla \times (\overline{E}^{i} + \overline{E}^{s}) = -j\omega\mu(\overline{H}^{i} + \overline{H}^{s}), \quad \nabla \times (\overline{H}^{i} + \overline{H}^{s}) = +j\omega\varepsilon(\overline{E}^{i} + \overline{E}^{s}). \tag{2.8}$$

The superscripts i and s are used to denote the incident and SFs, where the incident field is the field that would exist in the computational domain with no scatterers. If the computational domain is a free space, then the incident field satisfies Maxwell's equations, such that

$$\nabla \times \overline{E}^{i} = -j\omega\mu_0\,\overline{H}^{i}, \quad \nabla \times \overline{H}^{i} = +j\omega\varepsilon_0\,\overline{E}^{i}. \tag{2.9}$$

Substitution of (2.9) into (2.8) yields

$$\overline{E}^{s} = \frac{1}{j\omega\varepsilon}\nabla \times \overline{H}^{s} - \frac{(\varepsilon - \varepsilon_0)}{\varepsilon}\overline{E}^{i}, \tag{2.10}$$

$$\overline{H}^{s} = -\frac{1}{j\omega\mu}\nabla \times \overline{E}^{s} + \frac{(\mu_0 - \mu)}{\mu}\overline{H}^{i}. \tag{2.11}$$

In this chapter, the FDFD formulation for the 3D case is presented, and the method is applied to provide a solution for the subregions as well as for the entire domain for the sake of comparison. The field components of the incident plane wave for both θ and ϕ polarization can be given as

$$E_x^i(x,y) = E_\theta^i \cos\theta^i \cos\phi^i - E_\phi^i \sin\phi^i$$

$$E_x^i(x,y) = E_\theta^i \cos\theta^i \sin\phi^i + E_\phi^i \cos\phi^i$$

$$E_z^i(x,y) = -E_\theta^i \sin\theta^i$$

$$H_x^i(x,y) = \frac{1}{\eta_0}(E_\theta^i \cos\theta^i \cos\phi^i + E_\phi^i \sin\phi^i)$$

$$H_y^i(x,y) = \frac{1}{\eta_0}(E_\theta^i \cos\theta^i \sin\phi^i - E_\phi^i \cos\phi^i)$$

$$H_z^i(x,y) = \frac{1}{\eta_0}(-E_\phi^i \sin\theta^i) \tag{2.12}$$

where E_θ^i and E_ϕ^i indicate the polarization type, which can be written as

$$E_{\theta,\phi}^i = E_{\theta,\phi}^0\, e^{-j\beta(x\sin\theta^i \cos\phi^i + y\sin\theta^i + \sin\phi^i + z\cos\theta^i)} \tag{2.13}$$

where E_θ^0 and E_ϕ^0 are the magnitudes of the incident electric field, indicating whether the polarization is θ or ϕ polarized by assigning $E_\theta^0 = 1$ and $E_\phi^0 = 0$ or vice versa; k_o is the wave number; ε_o and μ_o are the permittivity and the permeability of the free space. The incident angle with respect to the x axis of the global coordinates system is ϕ, whereas θ^i is the incident angle with respect to the z axis. Having defined the incident fields, (2.10) can be written for the scattered electric field components in the form

$$E_x^s = \frac{1}{j\omega\varepsilon_{xy}} \frac{\partial H_z^s}{\partial y} - \frac{1}{j\omega\varepsilon_{xz}} \frac{\partial H_y^s}{\partial z} - \frac{(\varepsilon_{xi}-\varepsilon_0)}{\varepsilon_{xi}} E_x^i \qquad (2.14)$$

$$E_y^s = \frac{1}{j\omega\varepsilon_{yz}} \frac{\partial H_x^s}{\partial z} - \frac{1}{j\omega\varepsilon_{yx}} \frac{\partial H_z^s}{\partial x} - \frac{(\varepsilon_{yi}-\varepsilon_0)}{\varepsilon_{yi}} E_y^i \qquad (2.15)$$

$$E_z^s = \frac{1}{j\omega\varepsilon_{zx}} \frac{\partial H_y^s}{\partial x} - \frac{1}{j\omega\varepsilon_{zy}} \frac{\partial H_x^s}{\partial y} - \frac{(\varepsilon_{zi}-\varepsilon_0)}{\varepsilon_{zi}} E_z^i \qquad (2.16)$$

Using (2.11), the magnetic field components can be expressed in terms of the scattered electric field as

$$H_x^s = \frac{1}{j\omega\mu_{xz}} \frac{\partial E_y^s}{\partial z} - \frac{1}{j\omega\mu_{xy}} \frac{\partial E_z^s}{\partial y} + \frac{(\mu_o-\mu_{xi})}{\mu_{xi}} H_x^i \qquad (2.17)$$

$$H_y^s = \frac{1}{j\omega\mu_{yx}} \frac{\partial E_z^s}{\partial x} - \frac{1}{j\omega\mu_{yz}} \frac{\partial E_x^s}{\partial z} + \frac{(\mu_o-\mu_{yi})}{\mu_{yi}} H_y^i \qquad (2.18)$$

$$H_z^s = \frac{1}{j\omega\mu_{zy}} \frac{\partial E_x^s}{\partial y} - \frac{1}{j\omega\mu_{zx}} \frac{\partial E_y^s}{\partial x} + \frac{(\mu_o-\mu_{zi})}{\mu_{zi}} H_z^i. \qquad (2.19)$$

In (2.14–2.16) and (2.18–2.19), the permittivity and permeability parameters are indexed in such a way that these equations will be used for the PML region that will be used to truncate the computational domain and the non-PML regions as well. These parameters are defined in different ways depending on whether the node, at which the fields are to be evaluated, is inside the PML region or outside, such that, in the PML region,

$$\varepsilon_{xy} = \varepsilon_0 - j\frac{\sigma_y^e}{\omega}, \qquad \varepsilon_{yx} = \varepsilon_0 - j\frac{\sigma_x^e}{\omega}, \qquad \varepsilon_{xz} = \varepsilon_0 - j\frac{\sigma_z^e}{\omega}$$

$$\varepsilon_{zx} = \varepsilon_0 - j\frac{\sigma_x^e}{\omega}, \qquad \varepsilon_{zy} = \varepsilon_0 - j\frac{\sigma_y^e}{\omega}, \qquad \varepsilon_{yz} = \varepsilon_0 - j\frac{\sigma_z^e}{\omega}$$

$$\mu_{xy} = \mu_0 - j\frac{\sigma_y^m}{\omega}, \qquad \mu_{yx} = \mu_0 - j\frac{\sigma_x^m}{\omega}, \qquad \mu_{xz} = \mu_0 - j\frac{\sigma_z^m}{\omega}.$$

$$\mu_{zx}=\mu_0-j\frac{\sigma_x^m}{\omega}\ ,\qquad \mu_{zy}=\mu_0-j\frac{\sigma_y^m}{\omega}\ ,\qquad \mu_{yz}=\mu_0-j\frac{\sigma_z^m}{\omega}$$

$$\varepsilon_{xi}=\varepsilon_{yi}=\varepsilon_{zi}=\varepsilon_0 \qquad \mu_{xi}=\mu_{yi}=\mu_{zi}=\mu_0$$

Outside the PML region,

$$\varepsilon_{xy}=\varepsilon_{xi}=\varepsilon_x\ ,\quad \varepsilon_{yx}=\varepsilon_{yi}=\varepsilon_y\ ,\quad \varepsilon_{zx}=\varepsilon_{zi}=\varepsilon_z\ ,\quad \varepsilon_{xz}=\varepsilon_{xi}=\varepsilon_x$$
$$\varepsilon_{zy}=\varepsilon_{zi}=\varepsilon_z\ ,\quad \varepsilon_{yz}=\varepsilon_{yi}=\varepsilon_y$$
$$\mu_{xy}=\mu_{xi}=\mu_x\ ,\quad \mu_{yx}=\mu_{yi}=\mu_y\ ,\quad \mu_{zx}=\mu_{zi}=\mu_z\ ,\quad \mu_{xz}=\mu_{xi}=\mu_x$$
$$\mu_{zy}=\mu_{zi}=\mu_z\ ,\quad \mu_{yz}=\mu_{yi}=\mu_y$$

where σ_x^e, σ_y^e, σ_z^e, σ_x^m, σ_y^m and σ_z^m are the PML electric and magnetic conductivity distributions, and ε_x, ε_y, ε_z, μ_x, μ_y and μ_z are the anisotropic material parameters within the non-PML region of the computational space.

Equations (2.14–2.16) and (2.18–2.19) are the starting points to the construction of the FDFD equation for a 3D problem. Based on the Yee grid [26] shown in Figure 2.3, with only the field components associated with node (i, j, k) and after applying the central difference approximations to the derivatives in (2.14–2.16) and (2.18–2.19), we get

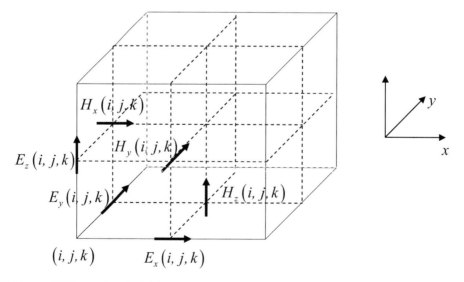

FIGURE 2.3: 3D Yee grid with field components related to node (i, j, k).

$$E_x^s(i,j,k) = \frac{1}{j\omega\varepsilon_{xy}(i,j,k)\,\Delta y}\,H_z^s(i,j,k) - \frac{1}{j\omega\varepsilon_{xy}(i,j,k)\,\Delta y}\,H_z^s(i,j-1,k)$$

$$-\frac{1}{j\omega\varepsilon_{xz}(i,j,k)\,\Delta z}\,H_y^s(i,j,k) + \frac{1}{j\omega\varepsilon_{xz}(i,j,k)\,\Delta z}\,H_y^s(i,j,k-1)$$

$$-\frac{(\varepsilon_{xi}(i,j,k)-\varepsilon_0)}{\varepsilon_{xi}(i,j,k)}\,E_x^i(i,j,k)$$

$$(2.20)$$

$$E_y^s(i,j,k) = \frac{1}{j\omega\varepsilon_{yz}(i,j,k)\,\Delta z}\,H_x^s(i,j,k) - \frac{1}{j\omega\varepsilon_{yz}(i,j,k)\,\Delta z}\,H_x^s(i,j,k-1)$$

$$-\frac{1}{j\omega\varepsilon_{yx}(i,j,k)\,\Delta x}\,H_z^s(i,j,k) + \frac{1}{j\omega\varepsilon_{yx}(i,j,k)\,\Delta x}\,H_z^s(i,j-1,k)$$

$$-\frac{(\varepsilon_{yi}(i,j,k)-\varepsilon_0)}{\varepsilon_{yi}(i,j,k)}E_y^i(i,j,k)$$

$$(2.21)$$

$$E_z^s(i,j,k) = \frac{1}{j\omega\varepsilon_{zx}(i,j,k)\,\Delta x}\,H_y^s(i,j,k) - \frac{1}{j\omega\varepsilon_{zx}(i,j,k)\,\Delta x}\,H_y^s(i-1,j,k)$$

$$-\frac{1}{j\omega\varepsilon_{zy}(i,j,k)\,\Delta y}\,H_x^s(i,j,k) + \frac{1}{j\omega\varepsilon_{zy}(i,j,k)\,\Delta y}\,H_x^s(i,j-1,k)$$

$$-\frac{(\varepsilon_{zi}(i,j,k)-\varepsilon_0)}{\varepsilon_{zi}(i,j,k)}E_z^i(i,j,k)$$

$$(2.22)$$

$$H_x^s(i,j,k) = \frac{1}{j\omega\mu_{xz}(i,j,k)\,\Delta z}\,E_y^s(i,j,k-1) - \frac{1}{j\omega\mu_{xz}(i,j,k)\,\Delta z}\,E_y^s(i,j,k)$$

$$-\frac{1}{j\omega\mu_{xy}(i,j,k)\,\Delta y}\,E_z^s(i,j+1,k) + \frac{1}{j\omega\mu_{xy}(i,j,k)\,\Delta y}\,E_z^s(i,j,k)$$

$$+\frac{(\mu_0-\mu_{xi}(i,j,k))}{\mu_{xi}(i,j,k)}H_x^i(i,j,k)$$

$$(2.23)$$

$$H_y^s(i,j,k) = \frac{1}{j\omega\mu_{yx}(i,j,k)\,\Delta x}\,E_z^s(i+1,j,k) - \frac{1}{j\omega\mu_{yx}(i,j,k)\,\Delta x}\,E_z^s(i,j,k)$$

$$-\frac{1}{j\omega\mu_{yz}(i,j,k)\,\Delta z}\,E_x^s(i,j,k) + \frac{1}{j\omega\mu_{yz}(i,j,k)\,\Delta z}\,E_x^s(i,j,k)$$

$$-\frac{\left(\mu_0 - \mu_{yi}(i,j,k)\right)}{\mu_{yi}(i,j,k)}H_y^i(i,j,k)$$

$$(2.24)$$

$$H_z^s(i,j,k) = \frac{1}{j\omega\mu_{zy}(i,j,k)\,\Delta y}\,E_x^s(i,j+1,k) - \frac{1}{j\omega\mu_{zy}(i,j,k)\,\Delta y}\,E_x^s(i,j,k)$$

$$-\frac{1}{j\omega\mu_{zx}(i,j,k)\,\Delta x}\,E_y^s(i+1,j,k) + \frac{1}{j\omega\mu_{zx}(i,j,k)\,\Delta x}\,E_y^s(i,j,k)$$

$$+\frac{\left(\mu_0 - \mu_{zi}(i,j,k)\right)}{\mu_{zi}(i,j,k)}H_z^i(i,j,k).$$

$$(2.25)$$

Equations (2.20–2.22) and (2.23–2.25) can be reduced to three equations, in terms of the three scattered electric field components. Thus, (2.20) can be rewritten as

$$E_x^s(i,j,k) = \frac{1}{j\omega\varepsilon_{xy}(i,j,k)\,\Delta y}\left(\begin{array}{l} \dfrac{E_x^s(i,j+1,k)}{j\omega\mu_{zy}(i,j,k)\Delta y} - \dfrac{E_x^s(i,j,k)}{j\omega\mu_{zy}(i,j,k)\Delta y} \\[2mm] -\dfrac{E_y^s(i+1,j,k)}{j\omega\mu_{zx}(i,j,k)\Delta x} \\[2mm] +\dfrac{E_y^s(i,j,k)}{j\omega\mu_{zx}(i,j,k)\Delta x} + \dfrac{\left(\mu_0 - \mu_{zi}(i,j,k)\right)}{\mu_{zi}(i,j,k)}H_z^i(i,j,k) \end{array} \right)$$

$$-\frac{1}{j\omega\varepsilon_{xy}(i,j,k)\,\Delta y}\left(\begin{array}{l} \dfrac{E_x^s(i,j,k)}{j\omega\mu_{zy}(i,j-1,k)\Delta y} - \dfrac{E_x^s(i,j-1,k)}{j\omega\mu_{zy}(i,j-1,k)\Delta y} \\[2mm] -\dfrac{E_y^s(i+1,j-1,k)}{j\omega\mu_{zx}(i,j-1,k)\Delta x} + \dfrac{E_y^s(i,j-1,k)}{j\omega\mu_{zx}(i,j-1,k)\Delta x} \\[2mm] +\dfrac{\left(\mu_0 - \mu_{zi}(i,j-1,k)\right)}{\mu_{zi}(i,j-1,k)}H_z^i(i,j-1,k) \end{array} \right)$$

$$-\frac{1}{j\omega\varepsilon_{xz}(i,j,k)\,\Delta z}\left(\begin{array}{c}\dfrac{E_z^{\,s}(i+1,j,k)}{j\omega\mu_{yx}(i,j,k)\,\Delta x}-\dfrac{E_z^{\,s}(i,j,k)}{j\omega\mu_{yx}(i,j,k)\,\Delta x}\\[2mm]-\dfrac{E_x^{\,s}(i,j,k+1)}{j\omega\mu_{yz}(i,j,k)\,\Delta z}+\dfrac{E_x^{\,s}(i,j,k)}{j\omega\mu_{yz}(i,j,k)\,\Delta z}\\[2mm]+\dfrac{\left(\mu_0-\mu_{yi}(i,j,k)\right)}{\mu_{yi}(i,j,k)}H_y^{\,i}(i,j,k)\end{array}\right)$$

$$+\frac{1}{j\omega\varepsilon_{xz}(i,j,k)\,\Delta z}\left(\begin{array}{c}\dfrac{1}{j\omega\mu_{yx}(i,j,k-1)\Delta x}E_z^{\,s}(i+1,j,k-1)\\[2mm]-\dfrac{1}{j\omega\mu_{yx}(i,j,k-1)\Delta x}E_z^{\,s}(i,j,k-1)\\[2mm]-\dfrac{1}{j\omega\mu_{yz}(i,j,k-1)\Delta z}E_x^{\,s}(i,j,k)\\[2mm]+\dfrac{1}{j\omega\mu_{yz}(i,j,k-1)\Delta z}E_x^{\,s}(i,j,k-1)\\[2mm]+\dfrac{\left(\mu_0-\mu_{yi}(i,j,k-1)\right)}{\mu_{yi}(i,j,k-1)}H_y^{\,i}(i,j,k-1)\end{array}\right)$$

$$-\frac{\left(\varepsilon_{xi}(i,j,k)-\varepsilon_0\right)}{\varepsilon_{xi}(i,j,k)}E_x^{\,i}(i,j,k).$$

(2.26)

Rewriting (2.26) will result into

$$E_x^{\,s}(i,j,k)=$$

$$+\left(\begin{array}{c}-\dfrac{E_x^{\,s}(i,j+1,k)}{\varepsilon_{xy}(i,j,k)\,\mu_{zy}(i,j,k)\omega^2\,(\Delta y)^2}+\dfrac{E_x^{\,s}(i,j,k)}{\varepsilon_{xy}(i,j,k)\,\mu_{zy}(i,j,k)\omega^2\,(\Delta y)^2}\\[2mm]+\dfrac{E_y^{\,s}(i+1,j,k)}{\varepsilon_{xy}(i,j,k)\,\mu_{zx}(i,j,k)\omega^2\Delta x\Delta y}-\dfrac{E_y^{\,s}(i,j,k)}{\varepsilon_{xy}(i,j,k)\mu_{zx}(i,j,k)\omega^2\Delta x\Delta y}\\[2mm]+\dfrac{\left(\mu_0-\mu_{zi}(i,j,k)\right)}{j\omega\Delta y\varepsilon_{xy}(i,j,k)\,\mu_{zi}(i,j,k)}H_z^{\,i}(i,j,k)\end{array}\right)$$

$$+\left(\begin{array}{l}\dfrac{E_x^s(i,j,k)}{\varepsilon_{xy}(i,j,k)\,\mu_{zy}(i,j-1,k)\,\omega^2(\Delta y)^2} \\[3mm] -\dfrac{E_x^s(i,j-1,k)}{\varepsilon_{xy}(i,j,k)\,\mu_{zy}(i,j-1,k)\,\omega^2(\Delta y)^2} \\[3mm] -\dfrac{E_y^s(i+1,j-1,k)}{\varepsilon_{xy}(i,j,k)\,\mu_{zx}(i,j-1,k)\,\omega^2\Delta x\Delta y}+\dfrac{E_y^s(i,j-1,k)}{\varepsilon_{xy}(i,j,k)\,\mu_{zx}(i,j-1,k)\,\omega^2\Delta x\Delta y} \\[3mm] -\dfrac{\left(\mu_0-\mu_{zi}(i,j-1,k)\right)}{j\omega\Delta y\varepsilon_{xy}(i,j,k)\,\mu_{zi}(i,j-1,k)}\,H_z^i(i,j-1,k)\end{array}\right)$$

$$+\left(\begin{array}{l}+\dfrac{E_z^s(i+1,j,k)}{\varepsilon_{xz}(i,j,k)\,\mu_{yx}(i,j,k)\,\omega^2\Delta x\Delta z}+\dfrac{E_z^s(i,j,k)}{\varepsilon_{xz}(i,j,k)\,\mu_{yx}(i,j,k)\,\omega^2\Delta x\Delta z} \\[3mm] -\dfrac{E_x^s(i,j,k+1)}{\varepsilon_{xz}(i,j,k)\,\mu_{yz}(i,j,k)\,\omega^2(\Delta z)^2} \\[3mm] +\dfrac{E_x^s(i,j,k)}{\varepsilon_{xz}(i,j,k)\,\mu_{yz}(i,j,k)\,\omega^2(\Delta z)^2} \\[3mm] -\dfrac{\left(\mu_0-\mu_{yi}(i,j,k)\right)}{j\omega\Delta z\varepsilon_{xz}(i,j,k)\,\mu_{yi}(i,j,k)}\,H_y^i(i,j,k)\end{array}\right)$$

$$+\left(\begin{array}{l}-\dfrac{E_z^s(i+1,j,k-1)}{\varepsilon_{xz}(i,j,k)\,\mu_{yx}(i,j,k-1)\,\omega^2\Delta x\Delta z}+\dfrac{E_z^s(i,j,k-1)}{\varepsilon_{xz}(i,j,k)\,\mu_{yx}(i,j,k-1)\,\omega^2\Delta x\Delta z} \\[3mm] +\dfrac{E_x^s(i,j,k)}{\varepsilon_{xz}(i,j,k)\,\mu_{yz}(i,j,k-1)\,\omega^2(\Delta z)^2}-\dfrac{E_x^s(i,j,k-1)}{\varepsilon_{xz}(i,j,k)\,\mu_{yz}(i,j,k-1)\,\omega^2(\Delta z)^2} \\[3mm] +\dfrac{\left(\mu_0-\mu_{yi}(i,j,k-1)\right)}{j\omega\Delta z\varepsilon_{xz}(i,j,k)\,\mu_{yi}(i,j,k-1)}\,H_y^i(i,j,k-1)\end{array}\right)$$

$$-\dfrac{\left(\varepsilon_{xi}(i,j,k)-\varepsilon_0\right)}{\varepsilon_{xi}(i,j,k)}E_x^i(i,j,k).$$

$$(2.27)$$

Finally, (2.20) can be written as

$$\dfrac{1}{\varepsilon_{xy}(i,j,k)\,\mu_{zy}(i,j,k)\,\omega^2(\Delta y)^2}E_x^s(i,j+1,k)$$

$$+\frac{1}{\varepsilon_{xy}(i,j,k)\,\mu_{zy}(i,j-1,k)\omega^2(\Delta y)^2}E_x^s(i,j-1,k)$$

$$+\frac{1}{\varepsilon_{xz}(i,j,k)\,\mu_{yz}(i,j,k)\omega^2(\Delta z)^2}E_x^s(i,j,k+1)$$

$$+\frac{1}{\varepsilon_{xz}(i,j,k)\,\mu_{yz}(i,j,k-1)\,\omega^2(\Delta z)^2}E_x^s(i,j,k-1)$$

$$+\left(\begin{array}{c}1-\dfrac{1}{\varepsilon_{xz}(i,j,k)\,\mu_{yz}(i,j,k-1)\,\omega^2(\Delta z)^2}-\dfrac{1}{\varepsilon_{xz}(i,j,k)\,\mu_{yz}(i,j,k)\,\omega^2(\Delta z)^2}\\[3mm]-\dfrac{1}{\varepsilon_{xy}(i,j,k)\,\mu_{zy}(i,j-1,k)\,\omega^2(\Delta y)^2}-\dfrac{1}{\varepsilon_{xy}(i,j,k)\,\mu_{zy}(i,j,k)\,\omega^2(\Delta y)^2}\end{array}\right)E_x^s(i,j,k)$$

$$-\frac{1}{\varepsilon_{xy}(i,j,k)\,\mu_{zx}(i,j,k)\omega^2\Delta x\Delta y}E_y^s(i+1,j,k)$$

$$+\frac{1}{\varepsilon_{xy}(i,j,k)\,\mu_{zx}(i,j-1,k)\omega^2\Delta x\Delta y}E_y^s(i+1,j-1,k)$$

$$-\frac{1}{\varepsilon_{xy}(i,j,k)\,\mu_{zx}(i,j-1,k)\omega^2\Delta x\Delta y}E_y^s(i,j-1,k)$$

$$+\frac{1}{\varepsilon_{xy}(i,j,k)\,\mu_{zx}(i,j,k)\,\omega^2\Delta x\Delta y}E_y^s(i,j,k)$$

$$+\frac{1}{\varepsilon_{xz}(i,j,k)\,\mu_{yx}(i,j,k-1)\omega^2\Delta x\Delta z}E_z^s(i+1,j,k-1)$$

$$-\frac{1}{\varepsilon_{xz}(i,j,k)\,\mu_{yx}(i,j,k)\,\omega^2\Delta x\Delta z}E_z^s(i+1,j,k)$$

$$-\frac{1}{\varepsilon_{xz}(i,j,k)\,\mu_{yx}(i,j,k-1)\omega^2\Delta x\Delta z}E_z^s(i,j,k-1)$$

$$+\frac{1}{\varepsilon_{xz}(i,j,k)\,\mu_{yx}(i,j,k)\omega^2\Delta x\Delta z}E_z^s(i,j,k)$$

$$=\frac{\left(\mu_0-\mu_{zi}(i,j,k)\right)}{j\omega\Delta y\varepsilon_{xy}(i,j,k)\,\mu_{zi}(i,j,k)}H_z^i(i,j,k)-\frac{\left(\mu_0-\mu_{zi}(i,j,k)\right)}{j\omega\Delta y\varepsilon_{xy}(i,j,k)\,\mu_{zi}(i,j-1,k)}H_z^i(i,j-1,k)$$

$$-\frac{\left(\mu_0-\mu_{yi}(i,j,k)\right)}{j\omega\Delta z\varepsilon_{xz}(i,j,k)\,\mu_{yi}(i,j,k)}H_y^i(i,j,k)+\frac{\left(\mu_0-\mu_{yi}(i,j,k-1)\right)}{j\omega\Delta z\varepsilon_{xz}(i,j,k)\,\mu_{yi}(i,j,k-1)}H_y^i(i,j,k-1)$$

$$-\frac{\left(\varepsilon_{xi}(i,j,k)-\varepsilon_0\right)}{\varepsilon_{xi}(i,j,k)}E_x^i(i,j,k)\ .$$

$$(2.28)$$

In the same manner, (2.21) takes the form

$$\frac{1}{\varepsilon_{yz}(i,j,k)\,\mu_{xz}(i,j,k)\omega^2(\Delta z)^2}E_y^s(i,j,k+1)$$

$$+\frac{1}{\varepsilon_{yz}(i,j,k)\,\mu_{xz}(i,j,k-1)\omega^2(\Delta z)^2}E_y^s(i,j,k-1)$$

$$+\frac{1}{\varepsilon_{yx}(i,j,k)\,\mu_{zx}(i,j,k)\omega^2(\Delta x)^2}E_y^s(i+1,j,k)$$

$$+\frac{1}{\varepsilon_{yx}(i,j,k)\,\mu_{zx}(i-1,j,k)\omega^2(\Delta x)^2}E_y^s(i-1,j,k)$$

$$+\left(\begin{array}{c}1-\dfrac{1}{\varepsilon_{yx}(i,j,k)\,\mu_{zx}(i-1,j,k)\,\omega^2(\Delta x)^2}-\dfrac{1}{\varepsilon_{yx}(i,j,k)\,\mu_{zx}(i,j,k)\omega^2(\Delta x)^2}\\[4mm]-\dfrac{1}{\varepsilon_{yz}(i,j,k)\,\mu_{xz}(i,j,k-1)\omega^2(\Delta z)^2}-\dfrac{1}{\varepsilon_{yz}(i,j,k)\,\mu_{xz}(i,j,k)\omega^2(\Delta z)^2}\end{array}\right)E_y^s(i,j,k)$$

$$-\frac{1}{\varepsilon_{yz}(i,j,k)\,\mu_{xy}(i,j,k)\omega^2\Delta y\,\Delta z}E_z^s(i,j+1,k)$$

$$+\frac{1}{\varepsilon_{yz}(i,j,k)\,\mu_{xy}(i,j,k-1)\omega^2\Delta y\,\Delta z}E_z^s(i,j+1,k-1)$$

$$-\frac{1}{\varepsilon_{yz}(i,j,k)\,\mu_{xy}(i,j,k-1)\omega^2\Delta y\,\Delta z}E_z^s(i,j,k-1)$$

$$+\frac{1}{\varepsilon_{yx}(i,j,k)\,\mu_{xy}(i,j,k)\omega^2\Delta y\,\Delta x}E_z^s(i,j,k)$$

$$+\frac{1}{\varepsilon_{yx}(i,j,k)\,\mu_{zy}(i-1,j,k)\omega^2\Delta y\,\Delta x}E_x^s(i-1,j+1,k)$$

$$-\frac{1}{\varepsilon_{yx}(i,j,k)\,\mu_{zy}(i,j,k)\omega^2\Delta y\,\Delta x}E_x^s(i,j+1,k)$$

$$-\frac{1}{\varepsilon_{yx}(i,j,k)\,\mu_{zy}(i-1,j,k)\omega^2 \Delta y\,\Delta x}E_x^s(i-1,j,k)$$

$$+\frac{1}{\varepsilon_{yx}(i,j,k)\,\mu_{zy}(i,j,k)\,\omega^2 \Delta y\,\Delta x}E_x^s(i,j,k)$$

$$=\frac{\left(\mu_0-\mu_{xi}(i,j,k)\right)}{j\omega\Delta z\varepsilon_{yz}(i,j,k)\,\mu_{xi}(i,j,k)}H_x^i(i,j,k)-\frac{\left(\mu_0-\mu_{xi}(i,j,k-1)\right)}{j\omega\Delta z\varepsilon_{yz}(i,j,k)\,\mu_{xi}(i,j,k-1)}H_x^i(i,j,k-1)$$

$$-\frac{\left(\mu_0-\mu_{zi}(i,j,k)\right)}{j\omega\Delta x\varepsilon_{yx}(i,j,k)\,\mu_{zi}(i,j,k)}H_z^i(i,j,k)+\frac{\left(\mu_0-\mu_{zi}(i-1,j,k)\right)}{j\omega\Delta x\varepsilon_{yx}(i,j,k)\,\mu_{zi}(i-1,j,k)}H_z^i(i-1,j,k)$$

$$-\frac{\left(\varepsilon_{yi}(i,j,k)-\varepsilon_0\right)}{\varepsilon_{yi}(i,j,k)}E_y^i(i,j,k)\,.$$

$$(2.29)$$

The same procedures are to be done for (2.22), which can be represented as

$$\frac{1}{\varepsilon_{zx}(i,j,k)\,\mu_{yx}(i,j,k)\omega^2(\Delta x)^2}E_z^s(i+1,j,k)$$

$$+\frac{1}{\varepsilon_{zx}(i,j,k)\,\mu_{yx}(i-1,j,k)\omega^2(\Delta x)^2}E_z^s(i-1,j,k)$$

$$+\frac{1}{\varepsilon_{zy}(i,j,k)\,\mu_{xy}(i,j,k)\omega^2(\Delta y)^2}E_z^s(i,j+1,k)$$

$$+\frac{1}{\varepsilon_{zy}(i,j,k)\,\mu_{xy}(i,j-1,k)\omega^2(\Delta y)^2}E_z^s(i,j-1,k)$$

$$+\left(\begin{array}{c}1-\dfrac{1}{\varepsilon_{zy}(i,j,k)\,\mu_{xy}(i,j-1,k)\,\omega^2(\Delta y)^2}-\dfrac{1}{\varepsilon_{zy}(i,j,k)\,\mu_{xy}(i,j,k)\omega^2(\Delta y)^2}\\[2ex]-\dfrac{1}{\varepsilon_{zx}(i,j,k)\,\mu_{yx}(i-1,j,k)\omega^2(\Delta x)^2}-\dfrac{1}{\varepsilon_{zx}(i,j,k)\,\mu_{yx}(i,j,k)\omega^2(\Delta x)^2}\end{array}\right)E_z^s(i,j,k)$$

$$-\frac{1}{\varepsilon_{zx}(i,j,k)\,\mu_{yz}(i,j,k)\omega^2\Delta z\,\Delta x}E_x^s(i,j,k+1)$$

$$+\frac{1}{\varepsilon_{zx}(i,j,k)\,\mu_{yz}(i-1,j,k)\omega^2\Delta z\,\Delta x}E_x^s(i-1,j,k+1)$$

$$-\frac{1}{\varepsilon_{zx}(i,j,k)\,\mu_{yz}(i-1,j,k)\,\omega^2\Delta z\,\Delta x}E_x^s(i-1,j,k)$$

$$+\frac{1}{\varepsilon_{zx}(i,j,k)\,\mu_{yz}(i,j,k)\,\omega^2\Delta z\,\Delta x}E_x^s(i,j,k)$$

$$+\frac{1}{\varepsilon_{zy}(i,j,k)\,\mu_{xz}(i,j-1,k)\,\omega^2\Delta z\,\Delta y}E_y^s(i,j-1,k+1)$$

$$-\frac{1}{\varepsilon_{zy}(i,j,k)\,\mu_{xz}(i,j,k)\,\omega^2\Delta z\,\Delta y}E_y^s(i,j,k+1)$$

$$-\frac{1}{\varepsilon_{zy}(i,j,k)\,\mu_{zx}(i,j-1,k)\,\omega^2\Delta z\,\Delta y}E_y^s(i,j-1,k)$$

$$+\frac{1}{\varepsilon_{zy}(i,j,k)\,\mu_{xz}(i,j,k)\,\omega^2\Delta z\,\Delta y}E_y^s(i,j-1,k)$$

$$=\frac{(\mu_0-\mu_{yi}(i,j,k))}{j\omega\Delta x\varepsilon_{zx}(i,j,k)\,\mu_{yi}(i,j,k)}H_y^i(i,j,k)-\frac{(\mu_0-\mu_{yi}(i-1,j,k))}{j\omega\Delta x\varepsilon_{zx}(i,j,k)\,\mu_{yi}(i-1,j,k)}H_y^i(i-1,j,k)$$

$$-\frac{(\mu_0-\mu_{xi}(i,j,k))}{j\omega\Delta y\varepsilon_{zy}(i,j,k)\,\mu_{xi}(i,j,k)}H_x^i(i,j,k)+\frac{(\mu_0-\mu_{xi}(i,j-1,k))}{j\omega\Delta x\varepsilon_{zy}(i,j,k)\,\mu_{xi}(i,j-1,k)}H_x^i(i,j-1,k)$$

$$-\frac{(\varepsilon_{zi}(i,j,k)-\varepsilon_0)}{\varepsilon_{zi}(i,j,k)}E_z^i(i,j,k).$$

$$(2.30)$$

A linear set of equations can be constructed using (2.28), (2.29), and (2.30). These equations can be arranged in a matrix form as $[A][E]=[Y]$, where $[A]$ is a $(3N\times3N)$ highly sparse coefficients matrix, $[E]$ is the unknown vector of length $(3N)$, in which the first (N) elements represent the E_x scattered electric field components, the second (N) elements represent the E_y scattered electric field components, and the third (N) elements represent the E_z scattered electric field components in the computation grid. The $[Y]$ is the excitation vector based on the right hand sides of (2.28), (2.29), and (2.30), respectively, and is a function of incident field components, E_x^i, E_y^i, E_z^i, H_x^i, H_y^i and H_z^i. The solution of this matrix equation for the vector $[E]$ yields the E_x^s, E_y^s, and E_z^s field components in the entire computational domain.

In this chapter, the constructed 3D code was developed using Fortran language due to the shortage of the current version of Matlab memory capabilities of addressing memory size larger than 2 GB, in addition to the faster computation using Fortran language relative to Matlab. Thus, a sparse matrix solver, based on the Fortran language, generated by [27], was used to efficiently solve this kind of sparse matrix equations, which also has the capabilities of reducing the matrix storage by keeping in memory only the nonzero elements of matrix [A].

2.4   PML ABSORBING BOUNDARY

To truncate the computational domain, layers of absorbing boundary based on the PML technique are used. The PML media properties surrounding the computational domain are chosen to effectively absorb all waves outgoing toward the boundaries. A perfectly conducting (PEC) layer is used to terminate the last layer of PML region. Theoretically, all waves are supposed to be sufficiently absorbed by the PML; thus, terminating the computational domain by a PEC layer will have no effect on the solution.

In the PML region, artificial conductivity is introduced such that it starts with very small value at the free-space PML interfaces with gradual increase until it reaches its maximum value at the last layer of the PML region. This maximum conductivity is denoted as σ_{max} and is defined as

$$\sigma_{max} = -\frac{\varepsilon_0 c\,(n)\ln\left[R\,(0)\right]}{2\delta_{PML}} \qquad (2.31)$$

where n is 0, 1, or 2 for a constant conductivity, linear conductivity, or a parabolic conductivity distribution, respectively. The parameter δ_{PML} is the PML thickness; c is the speed of light in free space, and $R(0)$ is the theoretical reflection factor on the free-space PML boundary for normal incidence. The conductivity distribution in the absorbing PML can be defined as

$$\sigma^e\,(h) = \sigma_{max}\left(\frac{h}{\delta_{PML}}\right)^n \qquad (2.32)$$

where h is the distance from the free-space PML interface to a specified PML. Electric and magnetic conductivities (σ^e and σ^m) need to be introduced in the PML region. The relation between these two conductivities can be defined as in [28] by

$$\frac{\sigma^e}{\varepsilon_0} = \frac{\sigma^m}{\mu_0}$$

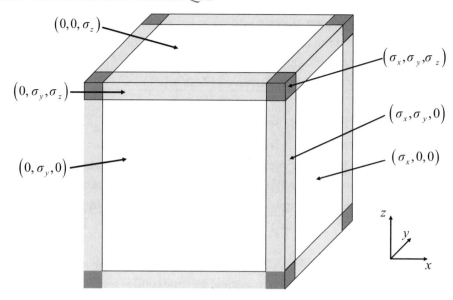

FIGURE 2.4: The computational domain with PML layers conductivity assignments.

Figure 2.4 shows the construction of the PML absorber just outside the original computational domain. It is clear from Figure 2.4 that only in the eight corners σ_x^{e}, σ_x^{m}, σ_y^{e}, σ_y^{m}, and σ_z^{e} σ_z^{m} are nonzeros. Other than that, the absorbing layers only have values for the conductivity in the x direction (i.e., $\sigma_x^{e} \neq 0$ and $\sigma_x^{m} \neq 0$). Similarly, the conductivity in the y direction (i.e., $\sigma_y^{e} \neq 0$ and $\sigma_y^{m} \neq 0$) as well as the conductivity in the z direction (i.e., $\sigma_z^{e} \neq 0$ and $\sigma_z^{m} \neq 0$).

•   •   •   •

CHAPTER 3

IMR Technique for Large-Scale Electromagnetic Scattering Problems: 3D Case

This chapter presents one of the methods to solve large electromagnetic problems, by dividing the computational domain into smaller subregions and solving each subregion separately. Then, the subregion solutions are combined to obtain a solution for the complete domain. An iterative approach using the FDFD method is presented to solve the scattering problem from 3D objects, with two different speeding up techniques used to efficiently solve the problems that can be divided into separated subregions. Some verification problems are presented to prove the validity of the presented technique in 3D objects.

3.1 ITERATIVE PROCEDURE BETWEEN MULTIPLE DOMAINS

The developed iterative technique is based on dividing the original electromagnetic scattering problem of a large domain into smaller problems in separated subregions, where the latest are to interact with each other to take into consideration the effect of the coupling between them. A huge saving in memory is achieved because dividing the original problem into smaller problems provides the benefit of minimizing the corresponding computational domain sizes, in addition to saving in the computational time, especially with large separation between some regions. Therefore, instead of dealing with one large region, which may not be possible in some problems because of limited computing resources, one would be dealing with multiple smaller regions.

The problem illustrated in Figure 3.1 shows the construction of the original domain, where the IMR technique is to be applied on. First, the original problem domain is divided into subregions and, in this case, three subregions. The scattered electromagnetic fields due to an incident wave are calculated separately in each subregion by the FDFD method. Then, fictitious electric and magnetic currents are calculated over imaginary surfaces surrounding the objects in each subregion, based on the equivalence principle. The electromagnetic fields radiated by these currents are calculated at the other subregions' grid nodes, using the formulation provided in [29] for 3D scatterers. These fields are considered as the new excitation for that region. As an example, for three objects, the fields exciting subregion 1 are a result of the superposition of the fields generated on subregion 1 using the

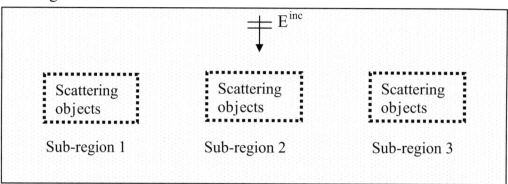

FIGURE 3.1: Original domain of a large scattering problem showing possible decomposition to two sub-domains.© 2006 IEEE.

currents calculated from subregions 2 and 3. Then the cycle of calculation of SFs, fictitious currents and radiated fields is repeated as a new iteration. The iteration process between regions continues until a convergence criterion is achieved. In our analysis, a stopping criterion of less than 2% difference in the calculated SFs at all angles from one iteration to the other is used. The sum of all calculated SFs through iterations gives the total SF, which is equivalent to the SF calculated from the solution of the original problem to a certain degree based on the total number of iterations. This iterative procedure is illustrated in Figure 3.2.

3.2   SPEEDING UP TECHNIQUES

It was realized that most of the consumed time by the subregion solution relative to the full domain solution is in the calculation of the field components generated due to the electric and magnetic currents from the opposing region. Thus, two approaches were studied and presented in this work to improve the timing issue: 1) interpolation process and 2) TF/SF technique.

1)   Interpolation process: An interpolation process was used in the field calculation, to speed up the process leading to 70% reduction in time with respect to the noninterpolated process. The interpolation process can be simply described as an averaging process, where just the odd plane cuts in the three directions (i.e., x, y, and z) are the only planes where the fields are to be computed in, whereas the even plane cuts are simply an average of the previous and following planes. Figure 3.3 describes the interpolation process, for the y direction, which is to be applied also for both the x and z directions to yield the 70% time reduction. The dotted planes are the averaged even planes, whereas the dashed planes are the planes where the fields are actually calculated based on the electric and magnetic currents generated from the opposite region.

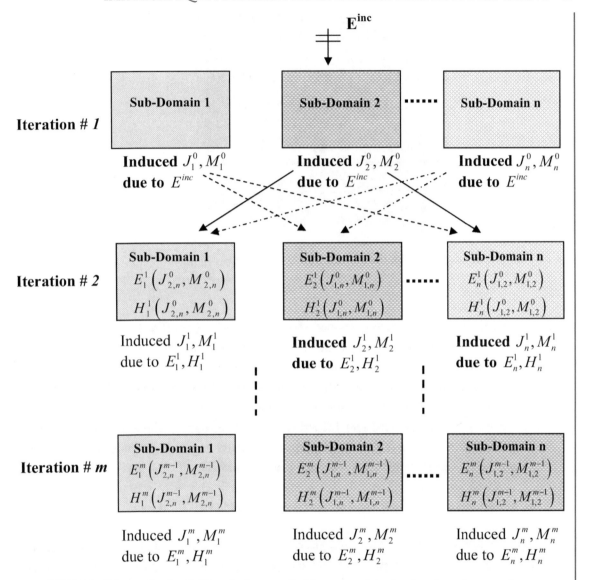

FIGURE 3.2: Scheme for the IMR technique applied by converting the electric and magnetic currents to field components generated on the other regions. © 2006 IEEE.

2) TF/SF technique: The TF/SF technique is combined with the FDFD method to speed up the calculations of the IMR technique, thus leading to a remarkable reduction in the time difference between the IMR solution and the full domain solution. The idea behind the time saving using the TF/SF technique is simply by calculating the fields on just two layers in the x, y, and z directions, instead of at each grid point in the entire subregion. Each computational domain of a subregion is divided into two regions separated by a nonphysical

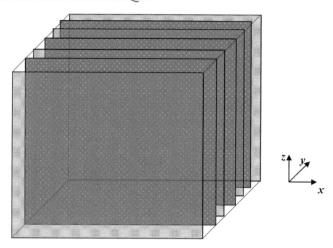

FIGURE 3.3: Configuration for the averaging process along the *y* direction.

virtual surface that is used to connect the fields in each region. The fields generated based on the computed electric and magnetic currents from the opposing subregion are generated on this virtual boundary where they act as an incident field on the same region. The TF/SF technique is used to solve for both the total and the SFs, where they are both considered as the unknown fields instead of using the classical SF approach for the entire subregion. Figure 3.4a illustrates the idea behind the TF/SF technique, where region 1, the inner zone, operates on the TF vector components, whereas region 2, the outer zone, operates on the SF vector components. For such reason, a problem of consistency arises when the required spatial differences are taken across the interface plane, as it would be inconsistent to perform an arithmetic operation between scattered and TF values.

For the ease of describing the technique, consider the 1D linear grid of E_z and H_y field components shown in Figure 3.4b. As shown before, the magnetic field components were eliminated by substituting them in the electric field expressions, as illustrated in the 3D FDFD Formulation section. Then, the Yee algorithm of the electric field is given at any grid point by

$$E_z(i) = \frac{1}{j\omega\varepsilon}\frac{1}{\Delta x}\Big[H_y(i+1/2)-H_y(i-1/2)\Big] \qquad (3.1)$$

which can then be rewritten as

$$E_z(i) = \frac{-1}{\omega^2\varepsilon\mu}\frac{1}{\Delta x^2}\Big[E_z(i+1)-2E_z(i)+E_z(i-1)\Big]. \qquad (3.2)$$

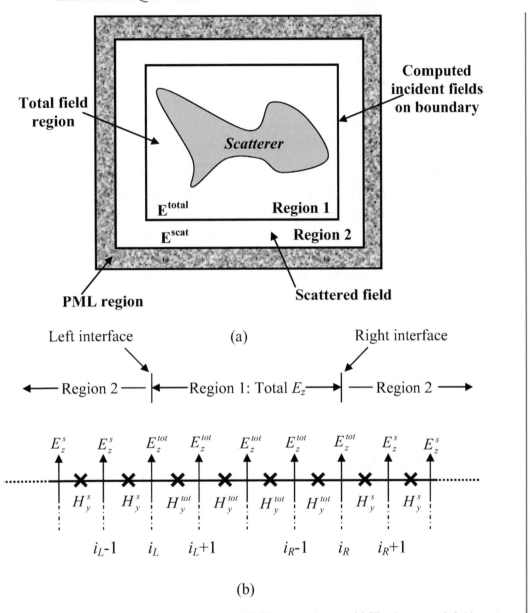

FIGURE 3.4: Total/scattered field zoning of the FDFD space lattice: (a) Total scattered field zoning connected by a virtual surface and absorbing boundary condition; (b) field component location in a one-dimensional (1D) *x*-directed cut. © 2006 IEEE.

This is valid when the two field components on the right-hand side are contained within a single grid region, whether SF or TF. Now let us consider applying (3.2) to the E_z component at the grid point i_L, located on the left interface between regions 1 and 2. Using (3.2) and generalizing the equation to the total electric field component yields

$$E_z^{\text{tot}}(i_L) = \frac{-1}{\omega^2 \varepsilon \mu} \frac{1}{\Delta x^2} \left[E_z^{\text{tot}}(i_L+1) - 2E_z^{\text{tot}}(i_L-1) \right]. \qquad (3.3)$$

One can notice that, in (3.3), the E_z component at nodes $i_L + 1$ and i_L-1 is defined as TF. Based on Figure 3.4a, which illustrates the idea of the TF/SF technique, the E_z component at node $i_L + 1$ defined at region 1 should be a TF. Although the E_z component at node i_L-1 defined at region 2 should be a SF. This means that the total electric field component at node i_L-1 should be replaced by the summation of the two field components, the scattered and the assumed known incident electric field components. This leads to a final representation of (3.3) as

$$E_z^{\text{tot}}(i_L) = \frac{-1}{\omega^2 \varepsilon \mu} \frac{1}{\Delta x^2} \left[E_z^{\text{tot}}(i_L+1) - 2E_z^{\text{tot}}(i_L) + E_z^{\text{s}}(i_L-1) \right] - \frac{1}{\omega^2 \varepsilon \mu \Delta x^2} E_z^{\text{i}}(i_L-1). \qquad (3.4)$$

Similar procedures are implemented to the numerical differentiation at the right interface between regions 1 and 2 (shown in Figure 3.4b).

In the same manner described for the 1D formulation, the two and three dimensions can be illustrated by applying the same procedures, which is explicitly described in [23].

3.3   NUMERICAL RESULTS

First, the radar cross-section (RCS) for some structures is computed to show the validity of the 3D FDFD code in solving any arbitrary shaped single domain problem, which for both a θ and ϕ polarized plane wave are expressed, respectively, in the following form

$$\sigma_{\theta\theta} = \lim_{r \to \infty} \left[4\pi r^2 \frac{|E_\theta^{\text{s}}|^2}{|E_\theta^{\text{i}}|^2} \right], \qquad \sigma_{\phi\phi} = \lim_{r \to \infty} \left[4\pi r^2 \frac{|E_\phi^{\text{s}}|^2}{|E_\phi^{\text{i}}|^2} \right]. \qquad (3.5)$$

Figure 3.5 shows the RCS of a conductor sphere excited by a θ polarized plane wave having $ka = 1.1$ and for incident angles $\phi^i = 0°$ and $\theta^i = 90°$, where the far field was calculated based on the computed magnetic and electric current densities over the artificial boundary surrounding the sphere [29]. Eight PML layers were used to terminate the computational domain with a reflection factor $R(0) = 1 \times 10^7$ and $n = 2$, considering a parabolic conductivity distribution. In addition to the eight PML, eight air buffer layers are introduced between the PML absorbing boundary and

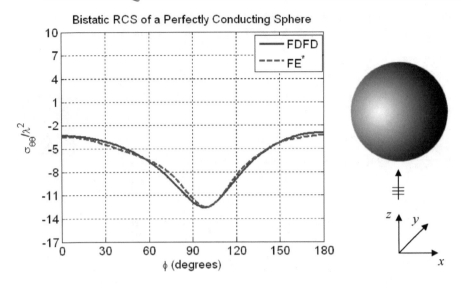

FIGURE 3.5: Bistatic RCS of a conductor sphere excited by a θ polarized plane wave for $\phi^i = 0°$ and $\theta^i = 90°$.

the simulated object. The number of layers simulating the PML and the air buffer, also the values of $R(0)$ and n, will all be the same in all the simulations and problems presented in this book, unless otherwise is mentioned. A discretization of 0.0175λ is used for all directions x, y, and z, leading to a computational domain size of 140,608 cells. The free-space wavelength (λ) is assumed to be equal to 1 m in all the simulated problems in this book, unless otherwise is mentioned. Excellent agreement between the FDFD computed results and those generated using the finite element method presented in [30] is observed.

Another verification to prove the validity of the developed 3D FDFD code is shown in Figure 3.6, where the RCS of a perfectly conducting cube of side length equal to 0.75λ excited by a θ polarized plane wave for $\phi^i = 90°$ and $\theta^i = 90°$ is presented. A discretization of 0.075λ is used in x, y, and z directions leading to a total computational size of 140,608 cells. The FDFD result shows very good agreement with the finite element results published in [30].

Figure 3.7 shows the geometry of two spheres, one having relative permittivity $\varepsilon_r = 3$, relative permeability $\mu_r = 1$, and radius $= 0.1\lambda$, and the other sphere has relative permittivity $\varepsilon_r = 1.001$, relative permeability $\mu_r = 1$, and radius $= 0.4\lambda$. The separation between the two spheres is 0.5λ. This configuration was constructed to check the validity of the IMR technique in comparison with that generated using the exact solution of a dielectric sphere having relative permittivity $\varepsilon_r = 3$ and radius equal to 0.1λ; this is the reason behind having a bigger sphere assigned a relative permittivity value

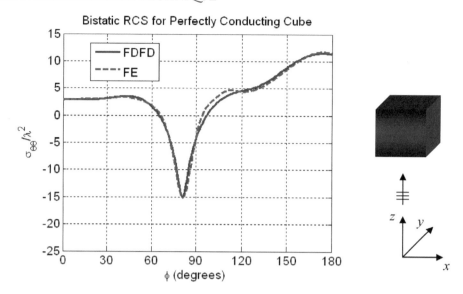

FIGURE 3.6: Bistatic RCS of perfectly conducting cube excited by a θ polarized plane wave for $\phi^i = 90°$ and $\theta^i = 90°$.

almost the same as free space. It is clear from the calculated RCS in Figure 3.8 based on the exact solution (full domain solution) and the subregion solution after two iterations that the results generated from the subregion solution converge to the exact solution, with very good agreement. For the full domain simulation of this problem, a discretization of 0.025λ was used in x, y, and z directions leading to a total computational size of 376,832 cells. For the IMR solution, the region simulating

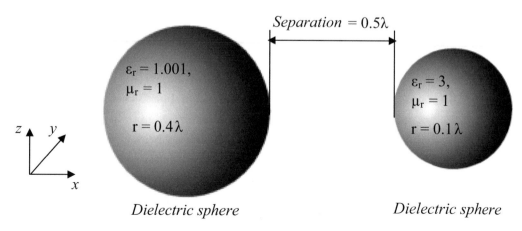

FIGURE 3.7: Geometry of two dielectric spheres.

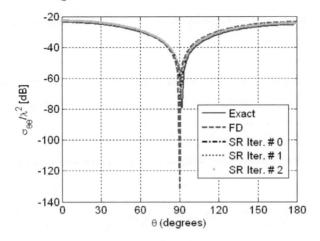

FIGURE 3.8: Bistatic RCS for the configuration defined in Figure 3.7. © 2006 IEEE.

the dielectric sphere of radius equal to 0.4λ was assigned a discretization value of 0.04λ in x, y, and z directions, whereas the region simulating the smaller dielectric sphere of radius equal to 0.1λ was assigned a discretization value of 0.02λ in all directions. This is one of the main features of the IMR technique in providing the flexibility of choosing different discretization in each subregion based on the simulated objects.

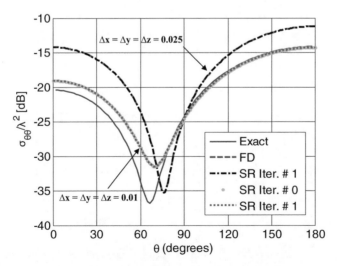

FIGURE 3.9: Bistatic RCS for a configuration similar to the one defined in Figure 3.7, but for a conductor sphere instead of the smaller dielectric sphere.

The same configuration illustrated in Figure 3.7 is used again, but this time, the dielectric sphere of relative permittivity equal to 3 is replaced by a conductor sphere. Figure 3.9 shows a comparison between the bistatic RCS of the exact solution for a conductor sphere, full domain solution, and subregion solution. It is obvious that when smaller discretization is used with the IMR technique, better results are achieved as can be seen in Figure 3.9. For this specific case, the use of smaller discretization prevented the full domain solution to be able to generate the results as it ran out of memory on the available 2 GB RAM 32-bit machine. On the other hand, using the IMR technique provides the flexibility of choosing different discretization for each subregion, based on

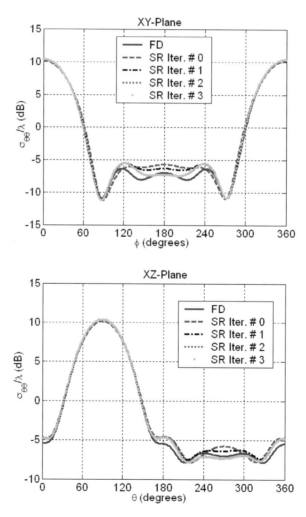

FIGURE 3.10: Bistatic RCS for both *xy* and *xz* plane cuts.

which, the smaller discretization was assigned to simulate the conductor sphere, whereas the larger discretization was used for the dielectric sphere (almost free space).

To verify the validity of the IMR technique described in this work, a problem of two dielectric spheres constructed in the same manner as the one described in Figure 3.7 is considered. The two spheres are separated by the same distance, which is 0.5λ. The sphere on the left hand side has a radius of 0.4λ, relative permittivity $\varepsilon_r = 2.2$, and relative permeability $\mu_r = 1$, whereas the sphere on the right hand side has a radius of 0.1λ, relative permittivity $\varepsilon_r = 3$, and relative permeability $\mu_r = 1$. Figure 3.10 shows the RCS calculated for the plane cuts xy and xz. The data for the third plane cut are very small and exactly over each other. The two spheres are excited by a θ polarized plane wave with $\phi^i = 0^0$ and $\theta^i = 90°$. The data in Figure 3.10 are computed using the FDFD solution based on a full domain simulation and compared with that generated using the IMR technique, using the FDFD method to solve for the problem in each subregion. A discretization of 0.02λ is used in the x, y, and z directions for the full domain simulation of this problem leading to a total number of cells of 554,688. As for the IMR solution, a discretization of 0.04λ was used to simulate the sphere of radius 0.4λ, and a discretization of 0.025λ was used in the other subregion containing

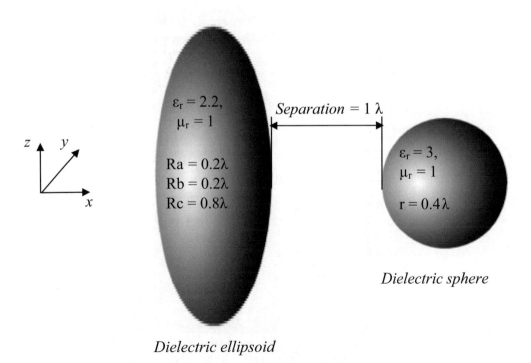

FIGURE 3.11: Geometry of a dielectric ellipsoid and a dielectric sphere.

the sphere of radius 0.1λ, leading to a total number of cells for both regions to be equal to 421,562 cells. The computational time of the full domain problem was recorded to be equal to 18 min 43 s, whereas that of the IMR solution after three iterations was 21 min 21 s. It can be seen that the IMR technique results converge to the full domain solution after three iterations, with a 24% memory reduction in the storage requirements and no significant change in the computational time due to the usage of the IMR technique with the aid of both the TF/SF approach and the flexibility of using different discretization in each of the domains relative to the original problem. The solution time would be even less, and the memory usage reduction would be more for configurations having larger separation between objects.

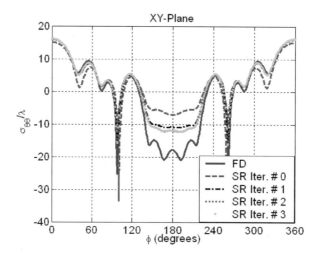

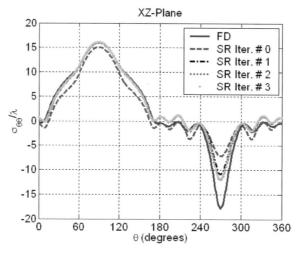

FIGURE 3.12: Bistatic RCS for both *xy* and *xz* plane cuts.

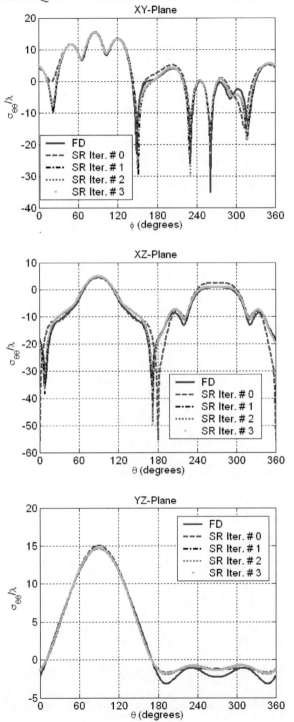

FIGURE 3.13: Bistatic RCS for *xy*, *xz*, and *yz* plane cuts.

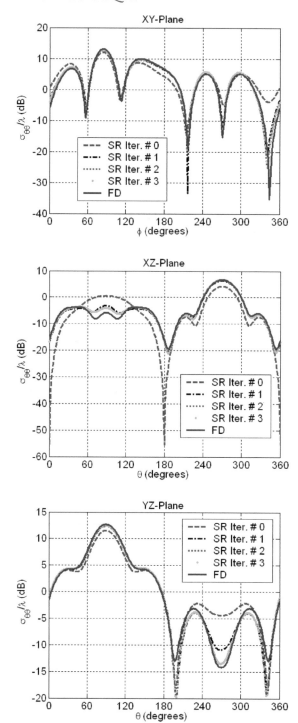

FIGURE 3.14: Bistatic RCS for *xy*, *xz*, and *yz* plane cuts. © 2006 IEEE.

Figure 3.11 shows another configuration that proves the validity of the IMR approach presented in this work for 3D structures. An ellipsoid with semiaxis; $Ra = 0.2\lambda$, $Rb = 0.2\lambda$, and $Rc = 0.8\lambda$, along the x, y, and z axes, respectively, is placed a distance of 1λ away from a sphere with radius equal to 0.4λ. The parameter λ is the free-space wavelength that is assumed to be equal to 1 m as previously mentioned. The ellipsoid has a relative permittivity $\varepsilon_r = 2.2$ and relative permeability $\mu_r = 1$, whereas the sphere on the right hand side has a relative permittivity $\varepsilon_r = 3$ and relative permeability $\mu_r = 1$. This configuration is excited by a θ polarized plane wave, with $\phi^i = 0^0$ and $\theta^i = 90°$. A discretization of 0.02λ is used in the x, y, and z directions for the full domain simulation of this problem leading to a total number of cells of 1,145,088. As for the IMR solution, a discretization of 0.02λ is used in the subregion simulating the ellipsoid, whereas a discretization of 0.04λ is used in the other subregion containing the sphere leading to a total number of cells for both regions to be equal to 443,456 cells. The computational time of the full domain problem is recorded to be equal to 38 min 54 s, whereas that of the IMR solution after three iterations was 36 min 32 s. Figure 3.12 shows the RCS calculated for the plane cuts xy and xz. It can be seen that the IMR technique results converge to the full domain solution after three iterations, with a 59% memory reduction in the storage requirements and no significant change in the computational time due to the usage of the IMR technique with the aid of both the TF/SF approach and the flexibility of using different discretization in each of the domains relative to the original problem. Using coarser discretization in one of the subregions relative to the full domain discretization is the reason behind the slight discrepancy shown between the IMR solution of the xy plane at 1802 and the full domain results. The same structure illustrated in Figure 3.11 is analyzed but with different incident angle ($\phi^i = 90^0$ and $\theta^i = 90°$). Figure 3.13 shows the RCS for the three plane cuts (i.e., xy, xz, and yz planes). Good agreement with the full domain solution is achieved for the three plane cuts after three iterations.

Simulating the same configuration presented in Figure 3.11, but instead of using a dielectric sphere, a conductor sphere is being used for a θ polarized excitation, having $\phi^i = 0^0$ and $\theta^i = 90°$.

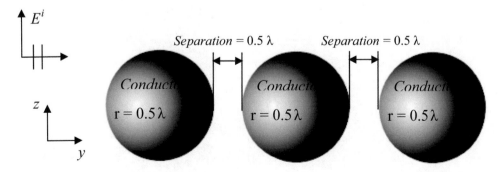

FIGURE 3.15: Geometry of three identical conducting spheres oriented along the y axis.

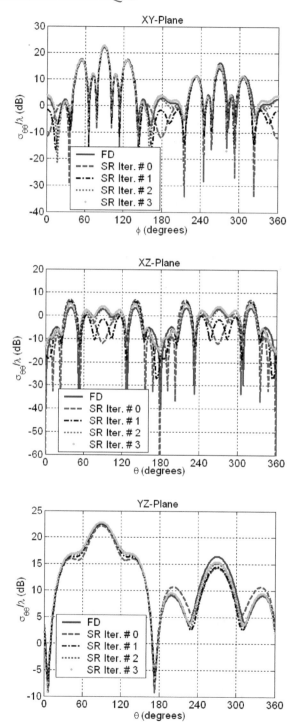

FIGURE 3.16: Bistatic RCS for *xy*, *xz*, and *yz* plane cuts. © 2006 IEEE.

The bistatic echo width is calculated and presented in Figure 3.14 for the three plane cuts, showing a good agreement between the results generated using the IMR technique with those generated using the full domain simulation.

For generality, a more complex configuration is presented in Figure 3.15 where three conducting spheres are excited by an incident plane wave with $\phi^i = 90°$ and $\theta^i = 90°$. The three spheres are oriented along the y axis, each having a radius of 0.5λ with a 0.5λ separation. The idea of this configuration is to prove the validity of the presented technique for more than two scattering objects or regions, where more interaction processes are required between the objects. The discretization used for both the full domain simulation as well as the IMR simulation of this problem is $\Delta x = \Delta y = \Delta z = 0.025\lambda$. For this specific configuration, the size of the computational domain for the full domain simulation is equal to 995,328 cells, whereas that of the IMR simulation for the three regions is equal to 1,119,744 cells. This is because, for this configuration, the separation between the three spheres in the full domain simulation will require less number of cells relative to the number of cells used to simulate the PML and air buffer layers in the IMR solution for all subregions. As stated earlier, the emphasis of this configuration is to prove the validity of the presented technique for simulating multiple objects specifically when the interaction process becomes more involved with small separation between the scatterers. Hence, in Figure 3.16, the RCS for the three plane cuts (i.e., xy, xz, and yz planes), is presented, proving the successful convergence of the data generated using the IMR technique to the full domain solution of this problem after three iterations. Good agreement between the results generated using the IMR technique with those generated using the full domain simulation can be recognized.

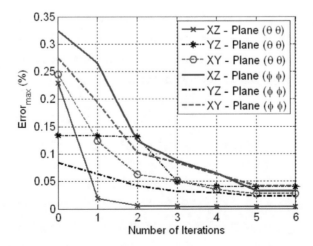

FIGURE 3.17: Maximum relative error for the problem illustrated in Figure 3.11 for both ($\theta\theta$) and ($\phi\phi$) polarizations. © 2006 IEEE.

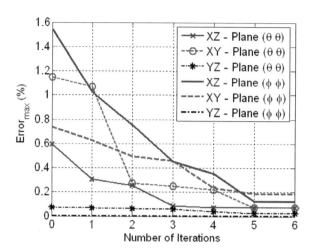

FIGURE 3.18: Maximum relative error for the problem illustrated in Figure 3.15, for both ($\theta\theta$) and ($\phi\phi$) polarizations. © 2006 IEEE.

To provide more information related to the convergence of the solution for the problem illustrated in Figure 3.11, but with a conductor sphere whose results are presented in Figure 3.14, and for the problem illustrated in Figure 3.15, for both ($\theta\theta$) and ($\phi\phi$) polarizations; a maximum relative error is defined at each iteration as

$$\left|\text{Error}\right|_{\max} = \max\left|\frac{\sigma_{\theta\theta/\phi\phi} - \sigma_{\theta\theta/\phi\phi}^{\text{FD}}}{\sigma_{\theta\theta/\phi\phi}^{\text{FD}}}\right| \times 100\% \, (\text{FD, finite difference}).$$

This error, for both problems, is plotted in Figures 3.17 and 3.18, respectively. It can be clearly noticed that as the number of iterations used within the IMR technique increases, the solution converges and reaches a stability condition, and the maximum relative error decreases. The level of the final error depends on the simulated configuration.

• • • •

CHAPTER 4

IMR Technique for Large-Scale Electromagnetic Scattering Problems: 2D Case

This chapter presents the use of the IMR algorithm to address 2D structures, where the electromagnetic fields radiated by the fictitious electric and magnetic currents calculated over the imaginary surfaces surrounding the objects in each subregion are calculated at the other subregions' grid nodes, using the formulation provided in Appendix 1. The FDFD solution for the 2D case is derived from the six FDFD field components of the 3D case, (2.20) and (2.21). This is the reason behind starting with the construction of the general 3D FDFD formulation in Chapter 2 so that one can easily descend to the special case of a 2D case as will be shown in this chapter. Some verification examples are presented to prove the validity of the presented technique for 2D scattering. PML, as an absorbing boundary, is used to terminate the computational domain.

4.1 2D FDFD FORMULATION

In this chapter, the FDFD formulation for the 2D TM_z case with $e^{j\omega t}$ time dependence is given, which can be easily extracted from the analysis provided for 3D structures presented in Chapter 2. The TM_z plane wave formulation for a 2D structure can be written as

$$E_z^i(x,y) = E_0\, e^{jk_0\left(x\cos\phi^i + y\sin\phi^i\right)}$$

$$H_z^i(x,y) = -\sin\phi^i \sqrt{\frac{\varepsilon_0}{\mu_0}}\, E_0\, e^{jk_0\left(x\cos\phi^i + y\sin\phi^i\right)}$$

$$H_y^i(x,y) = \cos\phi^i \sqrt{\frac{\varepsilon_0}{\mu_0}}\, E_0\, e^{jk_0\left(x\cos\phi^i + y\sin\phi^i\right)} \qquad (4.1)$$

where E_0 is the magnitude of the incident electric field; k_0 is the wave number; ε_0 and μ_0 and are the permittivity and the permeability of free space, respectively. The incident angle with respect to the x axis of the global coordinates system is ϕ^i.

In this chapter, the TM$_z$ plane wave is considered to excite 2D structures; this reduces the three field components E_x, E_y, and E_z in (2.28), (2.29), and (2.30) to just one field component (E_z) for 2D problems. Thus, (2.30) in Chapter 2 can be reduced to serve this purpose. Since only the z component of the scattered electric field is required, then by assigning E_y^s and E_x^s to be equal to zero (for the TM$_z$ case), we come up with the central finite difference approximation for E_z^s

$$E_{z(i,j)}^s - \frac{1}{j\omega(\Delta x)^2 \varepsilon_{zx(i,j)}} \left[\frac{1}{j\omega\mu_{yx\left(i+\frac{1}{2},j\right)}} \left(E_{z(i+1,j)}^s - E_{z(i,j)}^s \right) - \frac{1}{j\omega\mu_{yx\left(i+\frac{1}{2},j\right)}} \left(E_{z(i+1,j)}^s - E_{z(i,j)}^s \right) \right]$$

$$\frac{1}{j\omega(\Delta y)^2 \varepsilon_{zy(i,j)}} \left[\frac{-1}{j\omega\mu_{xy\left(i,j+\frac{1}{2}\right)}} \left(E_{z(i,j+1)}^s - E_{z(i,j)}^s \right) + \frac{1}{j\omega\mu_{xy\left(i,j-\frac{1}{2}\right)}} \left(E_{z(i,j)}^s - E_{z(i,j-1)}^s \right) \right]$$

$$= \left(\frac{\varepsilon_0}{\varepsilon_{zi(i,j)}} - 1 \right) E_{z(i,j)}^{inc} + \frac{1}{j\omega\Delta x \varepsilon_{zx(i,j)}} \left[\left(\frac{\mu_0}{\mu_{yi(i-1,j)}} - 1 \right) H_{y(i-1,j)}^{inc} - \left(\frac{\mu_0}{\mu_{yi(i,j)}} - 1 \right) H_{y(i,j)}^{inc} \right]$$

$$- \frac{1}{j\omega\Delta x \varepsilon_{zy(i,j)}} \left[\left(\frac{\mu_0}{\mu_{xi(i,j)}} - 1 \right) H_{x(i,j)}^{inc} - \left(\frac{\mu_0}{\mu_{xi(i,j-1)}} - 1 \right) H_{x(i,j-1)}^{inc} \right], \qquad (4.2)$$

where the permittivity and permeability parameters are indexed in the same manner as those in Chapter 2, to be used for the PML. Thus, for 2D structures, the media parameters to be assigned inside the PML region or outside are presented as follows.

In the PML region,

$$\varepsilon_{zx} = \varepsilon_0 - j\frac{\sigma_x^e}{\omega}, \qquad \mu_{yx} = \mu_0 - j\frac{\sigma_x^m}{\omega},$$

$$\varepsilon_{zy} = \varepsilon_0 - j\frac{\sigma_y^e}{\omega}, \qquad \mu_{xy} = \mu_0 - j\frac{\sigma_y^m}{\omega},$$

$$\varepsilon_i = \varepsilon_0 \qquad\qquad \mu_i = \mu_0,$$

Outside the PML region,

$$\varepsilon_{zi} = \varepsilon_{zx} = \varepsilon_{zy} = \varepsilon_z, \qquad \mu_{xi} = \mu_{xy} = \mu_x, \qquad \mu_{yi} = \mu_{yx} = \mu_y.$$

Equation (4.2) can be rewritten in the following general form as

$$a_{(i,j)} E_{(i-1,j)} + b_{(i,j)} E_{(i,j-1)} + c_{(i,j)} E_{(i,j)} + d_{(i,j)} E_{(i,j+1)} + e_{(i,j)} E_{(i+1,j)} = f_{(i,j)} \qquad (4.3)$$

where the subscript z and the superscript s are omitted for simplifying the presentation of the equation, and the coefficients a, b, c, d, and e are defined as

$$a = \frac{1}{(\Delta x)^2\,\omega^2\,\varepsilon_{zx(i,j)}\,\mu_{yx\left(i-\frac{1}{2},j\right)}}, \quad b = \frac{1}{(\Delta y)^2\,\omega^2\,\varepsilon_{zy(i,j)}\,\mu_{xy\left(i,j-\frac{1}{2}\right)}}, \quad d = \frac{1}{(\Delta y)^2\,\omega^2\,\varepsilon_{zy(i,j)}\,\mu_{xy\left(i,j+\frac{1}{2}\right)}}$$

$$e = \frac{1}{(\Delta x)^2\,\omega^2\,\varepsilon_{zx(i,j)}\,\mu_{yx\left(i+\frac{1}{2},j\right)}}, \quad c = 1 - a - b - d - e,$$

whereas the permeabilities at one-half cell away relative to the permittivity positions are given by

$$\mu_{yx\left(i-\frac{1}{2},j\right)} = \frac{1}{2}\left(\mu_{yx(i-1,j)} + \mu_{yx(i,j)}\right) \qquad \mu_{yx\left(i+\frac{1}{2},j\right)} = \frac{1}{2}\left(\mu_{yx(i+1,j)} + \mu_{yx(i,j)}\right)$$

$$\mu_{xy\left(i,j-\frac{1}{2}\right)} = \frac{1}{2}\left(\mu_{xy(i,j-1)} + \mu_{xy(i,j)}\right) \qquad \mu_{xy\left(i,j+\frac{1}{2}\right)} = \frac{1}{2}\left(\mu_{xy(i,j+1)} + \mu_{xy(i,j)}\right).$$

The electric and magnetic field components are defined on the 2D grid as shown in Figure 4.1. The magnetic field components are half a cell off grid from the nodes representing the electric field. As in Chapter 2, a linear set of equations can be constructed using (4.3) based on the node scheme given in Figure 4.1. These equations are arranged in the same matrix form, as $[A][E] = [Y]$, where $[A]$ in this case is an $(N \times N)$ highly sparse coefficients matrix, since we are just solving for one electric field component. The vector $[E]$ is the unknown E_z field values at all nodes in the domain, and $[Y]$ is the excitation vector representing the right-hand side of (4.3) and is a function of the incident field components E_z^i, H_x^i, and H_y^i.

The solution of this matrix equation for the vector $[E]$ yields the E_z^s field components in the computational domain. The sparsity pattern of $[A]$ can be observed as shown in Figure 4.2 for an example of (7×7) computational domain. A sparse matrix solver can be used to efficiently solve this kind of matrix equations. For the 2D analysis, the sparse matrix definition in Matlab is being used along with the command (sparse) to reduce the matrix storage by keeping in memory only the nonzero elements of matrix $[A]$.

For the TE$_z$ 2D configurations, one should start with the six field equations (2.20) and (2.21) to satisfy the TE$_z$ conditions. This should reduce to only one equation in the z component of the magnetic field.

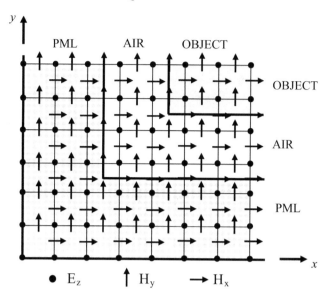

FIGURE 4.1: Node scheme for a TM$_z$ 2D scattering problem. Reproduced/modified by permission of American Geophysical Union.

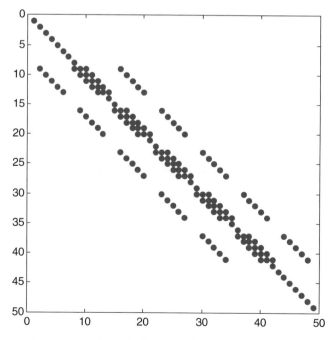

FIGURE 4.2: The sparsity pattern of matrix [A] for a (7×7). Reproduced/modified by permission of American Geophysical Union.

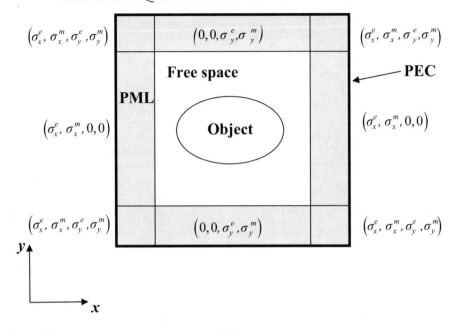

FIGURE 4.3: The computational domain with PML layers.

4.2   PML ABSORBING BOUNDARY

Similar to the analysis in Chapter 2, an absorbing boundary based on the PML technique is used, where the extended computational domain is terminated with a PEC. Figure 4.3 shows the construction of the PML absorber just outside the original computational domain. As it is clear from Figure 4.3, only in the four corners both σ_x^e, σ_x^m, and σ_y^e, σ_y^m are nonzeros. Other than that, the absorbing layers only have values for the conductivity in the x direction (i.e., $\sigma_x^e \neq 0$ and $\sigma_x^m \neq 0$, whereas $\sigma_y^e = 0$ and $\sigma_y^m = 0$). Similarly, the conductivity in the y direction is nonzero on both the upper and lower sides (i.e., $\sigma_x^e = 0$ and $\sigma_x^m = 0$, whereas $\sigma_y^e \neq 0$ and $\sigma_y^m \neq 0$).

4.3   NUMERICAL RESULTS

First, some results were computed to show the validity of the 2D FDFD code for solving any arbitrary-shaped 2D problem. Figure 4.4 shows the bistatic echo width of a PEC square excited by a TM$_z$ plane wave with $ks = 2$, for both $\phi^i = 45°$ and $\phi^i = 90°$, where the bistatic echo width can be calculated using the following formula,

$$\sigma_{2D} = \lim_{\rho \to \infty} \left[2\pi\rho \frac{\left| E_z^s \right|^2}{\left| E_z^i \right|^2} \right] \qquad (4.4)$$

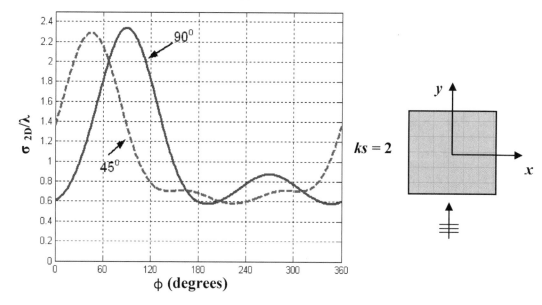

FIGURE 4.4: Bistatic echo width of a PEC square scatterer excited by a TM$_z$ plane wave for both $\phi^i = 45°$ and $\phi^i = 90°$.

A discretization of 0.03 in the x and y directions is used to simulate the problem described in Figure 4.4, with eight PML and eight air buffer layers as previously stated in Chapter 3. Excellent agreement between the FDFD computed results and the generated MoM results presented in [31] can be noticed.

Figure 4.5 shows the bistatic echo width of a thin PEC strip of $kL = 62$, where k is the wave number and L is the strip length. The strip is excited by a TMz plane wave at $\phi^i = 90°$. The purpose of the presented results in this case is to show another verification for the validity of the generated FDFD code. The generated FDFD result shows excellent agreement with those generated using the boundary value solution (BVS) reported in [32], where the BVS of the strip is obtained after simulating the strip by a linear array of parallel circular cylinders. A discretization of 0.2 in both directions is used.

This was the same for Figure 4.6, which shows a comparison between two different techniques, the FDFD and the BVS, to calculate the bistatic echo width of a circular conducting cylinder of radius 0.6λ excited by a TMz plane wave of $\phi^i = 180°$. For this configuration, a discretization of 0.06 is used to simulate the cylinder. The computed RCS shows excellent agreement between the two different techniques. Figure 4.7 also presents another comparison between the FDFD and the BVS for calculating the bistatic echo width of a rectangular dielectric scatterer of sides $kA = 12.57$ and $kB = 1.57$ having relative permittivity equal to 2 and relative permeability equal to 1.5, and

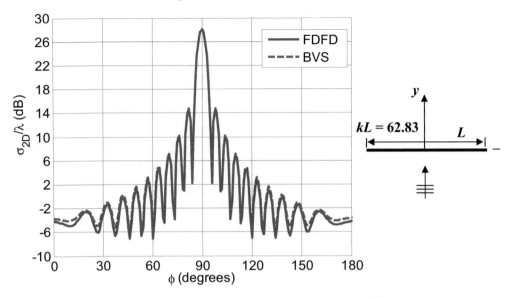

FIGURE 4.5: Comparison between the FDFD and the BVS for a 2D TM$_z$ bistatic scattering width from a thin conducting strip of length L = $62.83\lambda/2\pi = 10\lambda$.

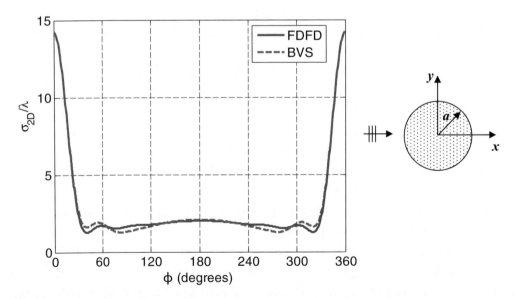

FIGURE 4.6: Comparison between the FDFD and the BVS for a 2D TM$_z$ bistatic scattering width from a circular conducting cylinder of radius $a = 0.6\lambda$.

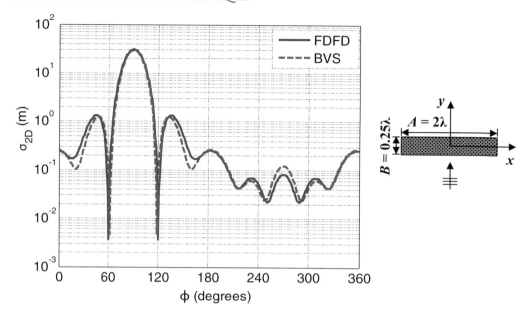

FIGURE 4.7: Comparison between the FDFD and the BVS for a 2D TM$_z$ bistatic scattering width from a dielectric rectangular cylinder.

excited by a TM$_z$ plane wave of $\phi^i = 2702$, using a discretization of 0.04 in both x and y directions. Excellent agreement between the two techniques can still be achieved, proving the validity of the generated 2D FDFD code.

After dealing with some cases to prove the validity of the FDFD technique, some numerical results are to be presented to show the solution of the IMR technique and its convergence to the full domain solution. Figure 4.8 shows the geometry of a 2D problem of one conducting and one dielectric cylinders of circular cross-section. The cylinders are separated by a distance 0.4λ along the x axis. The conductor cylinder has a radius of 0.1λ, whereas the dielectric cylinder radius is 0.5λ; the latter has relative permittivity $\varepsilon_r = 5$ and relative permeability $\mu_r = 1$. The incident electromagnetic plane wave field components on the cylinders are given by (4.1). The problem in Figure 4.8 is used to verify the validity of the IMR technique described in this work, because a rigorous solution of this problem can be obtained by the BVS [33].

Figure 4.9 shows the bistatic echo width calculated from the scattered E_z field component for the problem illustrated in Figure 4.8, where the two cylinders are excited by a TM$_z$ plane wave having $\phi^i = 90°$ from the positive x axis. The data in Figure 4.9 are computed using three different methods: first, the BVS [33]; second, the FDFD solution for the full domain; and third, the IMR procedure using FDFD method in each subregion. It can be seen that the IMR technique results

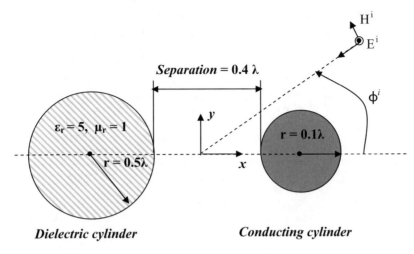

FIGURE 4.8: Geometry of one conducting and one dielectric cylinder and the direction of the incident plane wave. Reproduced/modified by permission of American Geophysical Union.

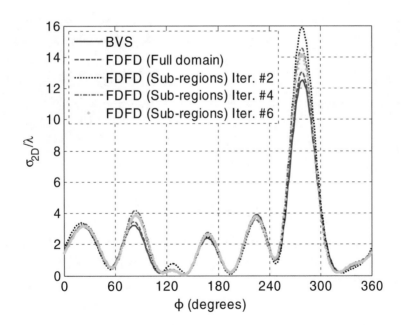

FIGURE 4.9: Comparison of the far field using BVS, FDFD for the full domain, and FDFD for subregions after 2, 4, and 6 iterations for the problem defined in Figure 4.8. Reproduced/modified by permission of American Geophysical Union.

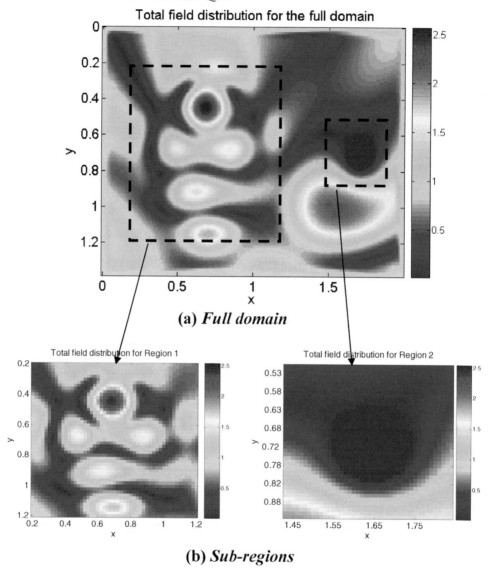

(a) Full domain

(b) Sub-regions

FIGURE 4.10: Comparison of the near-field distributions for the problem configuration presented in Figure 4.8 using FDFD for (a) full domain and (b) subregions. Reproduced/modified by permission of American Geophysical Union.

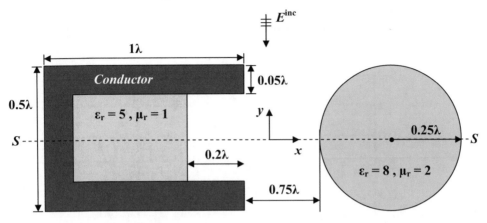

FIGURE 4.11: Geometry of the test configuration. Reproduced/modified by permission of American Geophysical Union.

converge to the full domain solution as the number of iteration increases. Because of the flexibility in discretizing the subregions, different cell sizes are used for the two subregions; the cell size used in the left subregion is 2 cm, whereas it is 1 cm in the right subregion. The cell size used in the full domain solution is 1 cm. The number of cells in the full domain is 27,800, whereas the total number of cells in the two subregions is 11,402. Therefore, 59% memory reduction in the storage requirements for this problem configuration is achieved by using the IMR technique. The computation time of the full domain solution is 1.30 min, whereas the total solution time of the two subregions is 1.55 min after four iterations, with the aid of the TF/SF defined in Chapter 3 computed using Matlab version 13 on a 1.9-GHz P4 personal computer.

Figure 4.10 shows a comparison between the near-field distributions generated using the FDFD code for the full domain simulation and the FDFD code for the two separate subregions using the IMR technique. The figures were generated for the subregions without the absorbing boundary (PML); therefore, they are compared with the parts that are surrounded by the dotted lines in the full domain plot.

A nonmagnetic rectangular dielectric cylinder having relative permittivity equal to 5 is embedded in a U-shaped conductor plate. A magnetic circular cylinder of relative permittivity equal to 8, relative permeability equal to 2, and radius equal to 0.25λ is placed at 0.75λ away from the first object, as shown in Figure 4.11. The cell size used for the full domain simulation is 1 cm, whereas those for the subregions are 1 and 1.25 cm for the first and second domains, respectively. Thus, the total size for the full domain problem is 23,496 cells, whereas for the two domains is 18,385 cells, which is 22% less than that of the full domain. This configuration is excited by a TM$_z$ plane wave

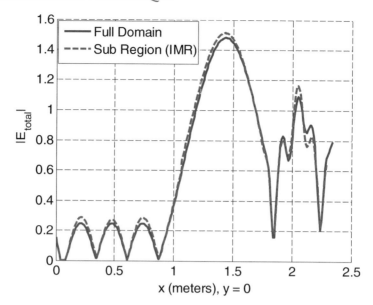

FIGURE 4.12: Comparison of the total near field generated along the *S-S* plane cut shown in Figure 4.11. Reproduced/modified by permission of American Geophysical Union.

with $\phi^i = 90°$. Shown in Figure 4.12 is a 2D plot representing a comparison between the total E_z field component on a cut along the center of the domain calculated from the full domain problem and the subregions solution. Although the IMR procedure does not deal with fields in the areas between the subregions, it is possible to compute the fields in these areas from the final currents evaluated in the separate regions. An example of this is shown in the middle area of Figure 4.12. Good agreement between the results can be recognized.

Figure 4.13 shows a comparison between the near-field distributions generated using the FDFD code for the full domain and the FDFD code for two separate subregions using the IMR technique. The figures generated for the subregions are without the absorbing boundary (PML), and they are to be compared with the parts that are surrounded by the dotted lines in the full do- main plot. Figure 4.14 shows a comparison of the bistatic echo width computed using the FDFD method for the full domain simulation with that generated using the IMR technique for the two subregions. Good agreement is achieved at all observation angles after 6, 18, and 26 iterations. Figure 4.14 shows the stability of the convergence of the solution after a large number of iterations. In both examples presented here, the number of iterations was determined based on the difference between the maximum values of the computed E_z in two successive iterations. A 1% difference is

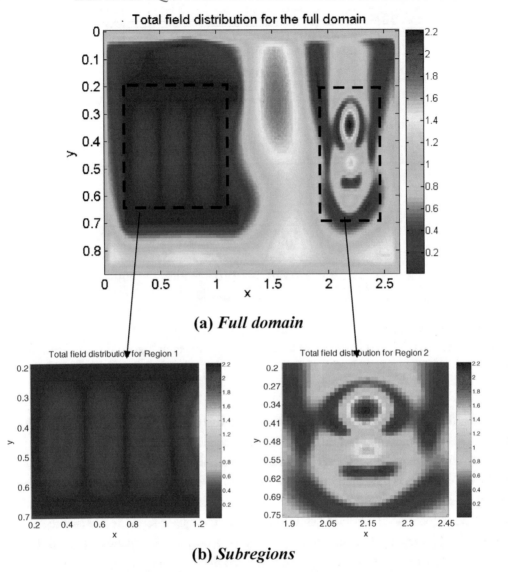

(a) *Full domain*

(b) *Subregions*

FIGURE 4.13: Comparison of the near-field distributions for the problem configuration presented in Figure 4.11 using FDFD for (a) full domain and (b) subregions. Reproduced/modified by permission of American Geophysical Union.

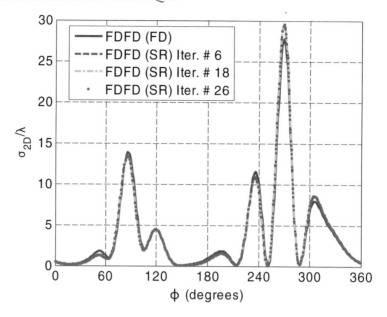

FIGURE 4.14: Comparisons of the far field using FDFD for the full domain and FDFD with IMR for subregions after 6, 8, and 26 iterations for the problem defined in Fig. 4.11. Reproduced/modified by permission of American Geophysical Union.

usually used to stop the iteration process; however, in this example, 26 iterations are used to confirm that this iterative solution is not diverging like other iterative techniques.

To show a more general overview on the interaction between multiple regions, the scattering from three cylinders is presented in terms of three separate subregions, and the results are compared with those generated using the BVS method for the three cylinders in one computational domain. As previously mentioned, the idea behind the IMR technique is to excite each subregion with the calculated fields generated because of the computed electric and magnetic currents from the other subregions. As an example, for the three objects, the fields exciting subregion 1 are a result of the superposition of the fields generated on subregion 1 using the currents calculated from subregions 2 and 3. This process is being repeated for the other subregions for several iterations, until the convergence criterion is being achieved. Figure 4.15 shows the configuration of the three dielectric cylinders having each relative permeability $\mu_r = 1$, relative permittivity $\varepsilon_r = 3$, and radius $= 0.5\lambda$. The cylinders are oriented along the x axis with a separation of 0.5λ. The three cylinders are excited by a TMz plane wave with $\phi^i = 90°$. The cell size used for the full domain simulation is 4 cm, whereas for each of the three subregions, the cell size is again 4 cm each. Thus, the total size of the full domain problem is 7,524 cells, whereas that for the three domains is 9,747 cells. Again, the emphasis of

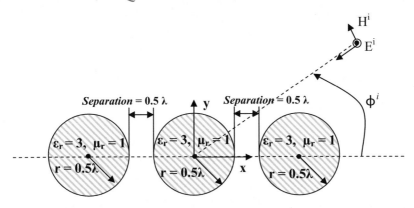

FIGURE 4.15: Geometry of three identical dielectric cylinders oriented along the x axis and the direction of the incident plane wave. Reproduced/modified by permission of American Geophysical Union.

this configuration is to show the performance of the IMR technique in providing an accurate solution for more than two scatterers when the interaction between them becomes more involved. The generated bistatic echo width for this configuration is shown in Figure 4.16, where the IMR results after four iterations using the FDFD method converge to the computed results generated using the BVS method. Good agreement between the two methods is recognized at all observation angles.

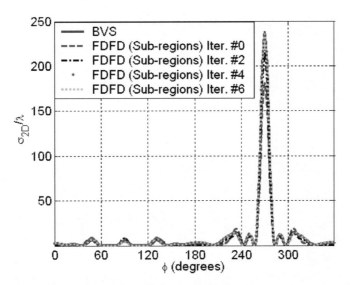

FIGURE 4.16: Comparisons of the far field using the FDFD with IMR for subregions after six iterations and the BVS method for the problem defined in Figure 4.15. Reproduced/modified by permission of American Geophysical Union.

CHAPTER 5

The IMR Algorithm Using a Hybrid FDFD and Method of Moments Techniques

This chapter presents a hybrid technique, which combines the desirable features of two different numerical methods, FDFD and MoM, to analyze large-scale electromagnetic problems by solving them individually and then applying the IMR technique. The FDFD/MoM approach takes advantage of the capability of the FDFD to analyze inhomogeneous bodies with arbitrary material properties and that of the MoM to model large metallic structures with less computational memory requirements. The presented hybrid technique in this chapter is being used to solve the 2D electromagnetic scattering problems, for which the RCS is calculated and compared with the regular FDFD/FDFD problem.

5.1   A 2D TM$_z$ EFIE–MoM FORMULATION FOR PEC CYLINDERS

MoM is widely used based on surface integral equation formulations, where unknowns (equivalent sources) are placed only over the surfaces/boundaries of homogeneous regions. These equivalent sources radiate into unbounded homogeneous space where their fields can be determined using the potential integrals, which in this chapter involves the 2D Green's function. To solve the problem of scattering from 2D perfectly electric conductor (PEC) cylinders, we will apply the equivalence principle by replacing the cylinder by an equivalent electric surface current density J that is placed on a contour C that represents the boundary of the cross-section of the conducting cylinder, as shown in Figure 5.1. The equivalent electric current density J on C is used to create zero fields internal to C where J is defined as

$$\bar{J} = \hat{n} \times \bar{H}. \tag{5.1}$$

The total tangential electric field in Figure 5.1 is continuous across C and equal to zero on C, which can be written as $\bar{E}_{\tan} = 0$. The incident fields are due to a 2D source distribution radiating

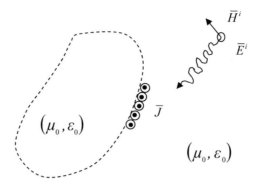

FIGURE 5.1: Scheme of the equivalence principle applied to simulate a PEC cylinder.

in homogeneous media and are thus continuous across C [34]. Thus, for a TM$_z$ plane wave, $\bar{E}_{\text{tan}} = 0$ can be rewritten as

$$\left[\bar{E}_z^i(\bar{r}) + \bar{E}_z^s(\bar{r}, \bar{J})\right]_{\text{tan}} = 0 \qquad \text{at } \rho \in C. \tag{5.2}$$

Since

$$\bar{E}_z^i = E_0 e^{jk_0(x\cos\phi^i\hat{x} + y\sin\phi^i\hat{y})}, \tag{5.3}$$

then from (5.2), the scattered electric field can be written as

$$\bar{E}_z^s = -\bar{E}_z^i. \tag{5.4}$$

Since

$$\bar{E}_z^s(\bar{r}, \bar{J}) = -j\omega\bar{A}(\bar{r}, \bar{J}) - \nabla\Phi\left(\bar{r}, \frac{\nabla'\cdot\bar{J}}{-j\omega}\right), \tag{5.5}$$

$$\bar{A}(\bar{r}, \bar{J}) = \mu_0 \int_c J_z(\ell')\hat{z}\left[\frac{1}{4j}H_o^{(2)}\left(k_0\left|\bar{\rho} - \bar{\rho}'\right|\right)\right]d\ell' \tag{5.6}$$

and

$$\nabla'\cdot\bar{J} = \frac{1}{\rho'}\frac{\partial}{\partial\rho'}\left(\rho'J_\rho\right) + \frac{1}{\rho'}\frac{\partial}{\partial\phi'}\left(J_\phi\right) + \frac{\partial}{\partial z'}\left(J_z\right) = 0 \tag{5.7}$$

then

$$\bar{E}_z^s(\bar{r}, \bar{J}) = -j\omega A_z(\bar{r}, \bar{J})\hat{z}$$

$$= -j\omega\mu_0 \int_c J_z(\ell')\left[\frac{1}{4j}H_0^{(2)}\left(k_0\left|\bar{\rho} - \bar{\rho}'\right|\right)\right]\partial\ell'\hat{z} \tag{5.8}$$

where $H_0^{(2)}$ is the Hankel function of the second kind and 0th order, $\bar{\rho}$ and $\bar{\rho}'$ are the position vectors for the observation and source points, respectively. Substituting (5.3) and (5.8) into (5.4), we get

$$-\frac{\omega\mu_0}{4} \int_c J_z\left(\ell'\right) H_0^{(2)} \left(k_0 \left|\bar{\rho} - \bar{\rho}'\right|\right) \partial\ell' = -E_0\, e^{jk_0\left(x\cos\phi^i\hat{x} + y\sin\phi^i\hat{y}\right)}. \tag{5.9}$$

Expanding the unknown current component, $J_z\left(\ell'\right)$ and using the point matching technique, the integral equation in (5.9) reduces to [29]

$$-\frac{\omega\mu_0}{4} \int_c \sum_{n=1}^N I_n \Pi_n\left(\ell'\right) H_0^{(2)} \left(k_0 \left|\bar{\rho}_m - \bar{\rho}'\right|\right) \partial\ell' =$$
$$-E_0\, e^{jk_0\left(x_m\cos\phi^i\hat{x} + y_m\sin\phi^i\hat{y}\right)}, \quad m = 1, 2, \ldots, N \tag{5.10}$$

where ρ_m is the position vector of the matching points and $\left\{\Pi_n\left(\ell'\right), n = 1, 2, \ldots, N\right\}$ is the set of pulse basis functions defined on the segments ΔC_n of the contour C by

$$\Pi_n(\rho) = \begin{cases} 1, & \rho \in \Delta C_n \\ 0, & \text{otherwise}. \end{cases} \tag{5.11}$$

Equation (5.10) may be written in the matrix form

$$[Z_{mn}][I_n] = [V_m] \tag{5.12}$$

where

$$V_m = -E_z^i\left(\bar{\rho}_m\right) = -E_0\, e^{jk_0\left(x_m\cos\phi^i\hat{x} + y_m\sin\phi^i\hat{y}\right)} \tag{5.13}$$

$$Z_{mn} = -\frac{\omega\mu_0}{4} \int_{\Delta c_n} H_0^{(2)} \left(k_0 \left|\bar{\rho}_m - \bar{\rho}'\right|\right) \partial\ell'. \tag{5.14}$$

Sample computed echo widths of different scatterers are provided to prove the validity of the developed MoM code for analyzing 2D PEC structures as shown in Figures 5.2 and 5.3. Figure 5.2 shows the bistatic echo width of a conducting cylinder of radius 0.6λ, excited by a TM_z plane wave incident at $\phi^i = 180°$. Excellent agreement can be realized from the figure shown for the computed data using the MoM technique analyzed in this chapter and the BVS [32]. Figure 5.3 shows the bistatic echo width of a thin PEC strip of $kL = 62$, where k is the wave number and L is the strip length. The strip is excited by a TM_z plane wave incident at $\phi^i = 90°$. The presented results are to show another verification for the validity of the generated MoM code. The generated MoM results

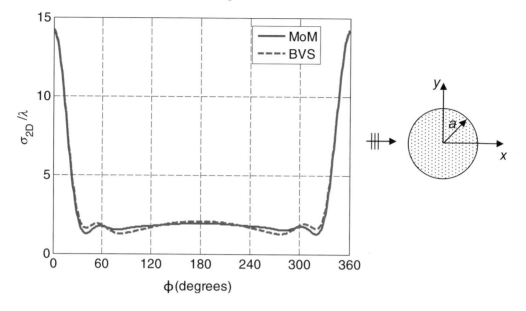

FIGURE 5.2: Comparison between the MoM and the BVS for a 2D TM$_z$ bistatic scattering width from a circular conducting cylinder.

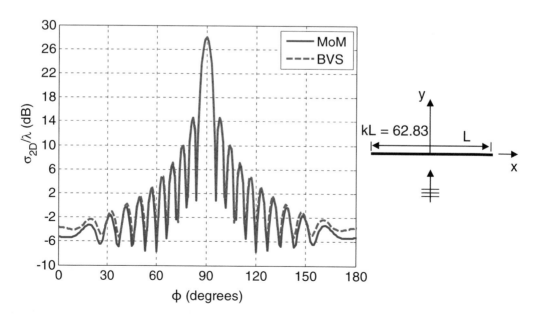

FIGURE 5.3: Comparison between the MoM and the BVS for a 2D TM$_z$ bistatic scattering width from a thin conducting strip.

show excellent agreement with that based on the BVS [32], where the BVS of the strip is obtained after simulating the strip by a linear array of parallel circular cylinders.

5.2   HYBRID FDFD/MoM TECHNIQUE

Hybrid methods, which combine the desirable features of two or more techniques, have been developed previously to analyze complex electromagnetic problems [35–41]. In this chapter, a hybrid

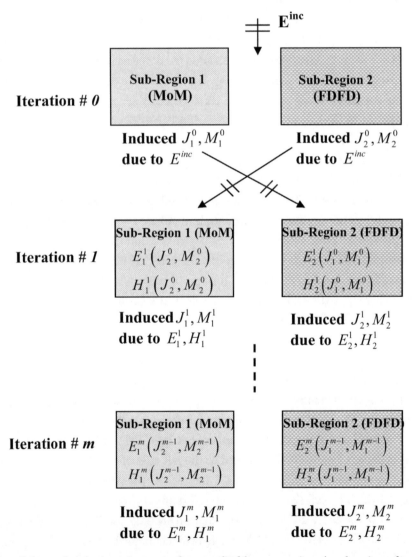

FIGURE 5.4: Scheme for the iterative procedure applied by converting the electric and magnetic currents to field components generated on the other region. © 2006 PIER.

FDFD/MoM approach is presented, which uses the advantage of the capability of the FDFD to analyze inhomogeneous bodies with arbitrary material properties and that of the MoM to model large metallic structures with less computational memory requirements. Thus, less computational time and memory consumption can be achieved.

Consider a problem of two scatterers apart from each other by a distance equivalent to the wavelength λ. The first step is to divide the original computational domain problem into subregions, namely in this case region 1 and region 2 and assume that the analysis for region 1 is performed using the MoM solution, whereas that of region 2 is performed using the FDFD method. Because in this chapter the MoM is used to simulate conducting structures only, electric current is calculated due to an incident plane wave excitation based on (5.12), whereas the scattered electromagnetic fields due to an incident wave are calculated from region 2 (FDFD solution). Based on the equivalence principle, fictitious electric and magnetic currents are calculated over imaginary surfaces surrounding the scatterers of region 2, where these currents are used to generate electromagnetic fields at the positions of the excitation vector in region 1 (MoM solution). Once the electric current is computed using the MoM solution for region 1, electromagnetic fields radiated by this current are calculated over an imaginary surface where new electric and magnetic currents are generated. Electromagnetic fields radiated by these currents are then calculated at the other subregions' grid nodes for region 2 (FDFD solution). These fields are considered as the new excitation for that region, and the cycle of calculation of SFs, fictitious currents, and radiated fields is repeated as a new iteration. The iteration process between subregions continues until a convergence criterion is achieved. The sum of all calculated SFs through iterations gives the total SF, which is found to be equivalent to the SF calculated from the solution of the original problem with acceptable tolerance. This iterative procedure involving the IMR technique, where a hybrid solution of MoM and FDFD methods takes place, is illustrated in Figure 5.4.

5.3   NUMERICAL RESULTS

In this section, numerical results are provided to prove the validity of this hybridization method (FDFD/MoM) using the IMR technique. Figure 5.5 shows a test configuration to prove such idea, where two cylinders placed along the x axis are excited by a TM_z plane wave incident at $\phi^i = 180°$. A conducting cylinder of radius 0.5λ is placed at the left side of a dielectric cylinder of radius 1λ, where the latter has relative permittivity $\varepsilon_r = 2.2$ and relative permeability $\mu_r = 1$. The two cylinders are separated by 0.5λ. The cell size used for the full domain simulation is 4.25 cm, whereas those for the subregions of the IMR–FDFD/FDFD solution are 4.25 and 7.5 cm for the first and second domains, respectively. The IMR–FDFD/MoM hybrid technique is used to solve the same problem where the MoM is used to solve the SF from the conducting circular cylinder, whereas the FDFD

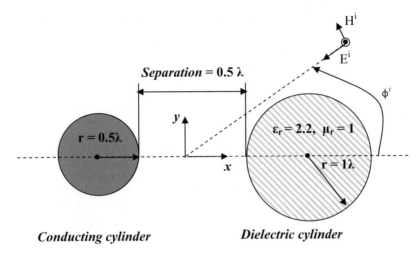

FIGURE 5.5: Geometry of one conducting and one dielectric cylinder and the incident plane wave direction. © 2006 PIER.

method is used to solve that of the dielectric cylinder. Hence, for the IMR–FDFD/MoM solution, the cell size used in the subregion enclosing the dielectric cylinder, solved using the FDFD method, is 7.5 cm, whereas the number of segments used for the conducting cylinder solved using the MoM is 250 segments.

Once the currents are generated, the interaction between the two cylinders takes place by applying the IMR technique as described in Section 5.2. One of the advantages of using the MoM technique is to reduce the computational time required to solve the conducting structures; thus, the consumed time by the hybrid IMR–FDFD/MoM solution after some iteration is expected to be less than that used for the same number of iterations by the IMR–FDFD/FDFD solution.

Figure 5.6 shows the far field calculations for the problem defined in Figure 5.5, where a comparison between three different approaches is presented: the full domain solution based on FDFD, the IMR–FDFD/FDFD solution and the hybrid IMR–FDFD/MoM solution. It can be clearly seen from the figure the strong match between the FDFD/FDFD and the hybrid FDFD/MoM solutions after four iterations, where both solutions approach the full domain solution using the FDFD technique. Table 5.1 shows a comparison between the three approaches, regarding the total computational domain size, which is involved in constructing the matrix solution, as well as the total computational time for each problem. Table 5.1 indicates that the total computational size using the IMR–FDFD/FDFD technique, for the problem defined in Figure 5.5, is 27% less than solving the classical FDFD solution applied to the whole problem; this is because of the flexibility in discretizing each domain separately and, thus, no obligation on using a smaller discretization when

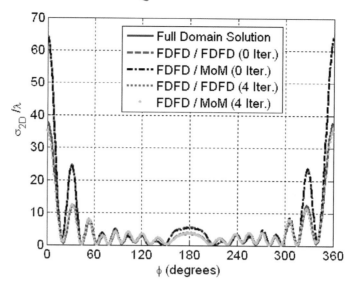

FIGURE 5.6: Bistatic echo width a conducting and dielectric cylinder excited by a TM$_z$ plane wave incident at $\phi^i = 180°$. © 2006 PIER.

defining a simple structure. The required computational size for the hybrid IMR–FDFD/MoM technique is 60% less than that required to solve the whole problem, which proves the efficiency of using the hybrid FDFD/MoM solution together with the IMR technique, regarding the computational memory consumption. Concerning the computational time, the hybrid IMR–FDFD/MoM converged to the full domain solution after four iterations in 38 s, whereas the IMR–FDFD/FDFD converged after the same number of iterations in 48 s, both compared with the full domain solution that took 36 s on a P4 machine with a 3.2-GHz processor and 2-GB RAM.

Figure 5.7 shows another configuration that consists of two different scatterers, a conducting square of side 0.5λ and a dielectric ellipse of relative permittivity equal to 2.2 and relative permeability

TABLE 5.1: Comparison among the full domain solution, the IMR–FDFD/FDFD, and the IMR–FDFD/MoM techniques regarding both the computational domain size and the computational time for the problem illustrated in Figure 5.5

	FULL DOMAIN	IMR–FDFD/FDFD	IMR–FDFD/MOM
Total domain size	9006 (cells)	6617 (cells)	3731 (cells)
Computational time (s)	36	48	38

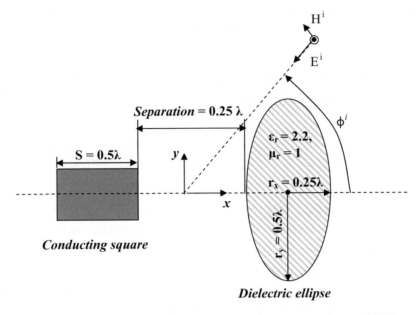

FIGURE 5.7: Geometry of a conducting square and a dielectric ellipse. © 2006 PIER.

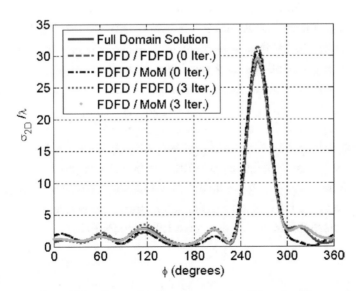

FIGURE 5.8: Bistatic echo width of a conducting square and a dielectric ellipse excited by a TM_z plane wave incident at $\phi^i = 90°$. © 2006 PIER.

TABLE 5.2: Comparison between the full domain solution, the IMR–FDFD/FDFD and the IMR–FDFD/MoM techniques regarding both the computational domain size and the computational time for the problem illustrated in Figure 5.7

	FULL DOMAIN	IMR–FDFD/FDFD	IMR–FDFD/MOM
Total domain size	14 784 (cells)	7749 (cells)	4214 (cells)
Computational time (s)	39	48	40

equal to 1. The radius of the ellipse in the x direction is equal to 0.25λ, and in the y direction is 0.5λ, the separation between the two scatterers is 0.25λ. The proposed structure in Figure 5.7 is excited by a TM_z plane wave incident at $\phi^i = 90°$. For this configuration, the cell size used for the full domain is 1.25 cm, leading to a total number of cells to be 14,784. As for the IMR–FDFD/FDFD solution, the cell size used in each of the domains is 1.25 and 4.0 cm for the first and second domains, respectively. The IMR–FDFD/MoM hybrid technique is used to solve the same problem where the MoM is used to solve the SF from the conducting square, whereas the FDFD method is used to solve that of the dielectric ellipse. Hence, for the IMR–FDFD/MoM solution, the cell size used in the subregion enclosing the dielectric ellipse, solved using the FDFD method, is 2.0 cm, whereas the number of segments used for the conducting cylinder solved using the MoM is 470 segments.

Figure 5.8 shows a comparison between the bistatic echo width of the three approaches, as presented previously in Figure 5.6. Again, in the hybrid IMR–FDFD/MoM solution, the MoM solution is used to solve the SF from the conducting square, whereas the FDFD is used to solve that of the dielectric ellipse. Still, a strong match can be noticed between the FDFD/FDFD and the hybrid FDFD/MoM solutions at zero iterations and after three iterations, where both solutions approach the full domain solution. Table 5.2 shows the total computational size using the IMR–FDFD/FDFD technique, for the problem defined in Figure 5.7, to be 50% less than solving the classical FDFD solution applied to the full problem, whereas for the hybrid IMR–FDFD/MoM technique, the required computational size was 70% less than that required to solve the full problem. The full domain solution required 39 s to finalize the simulation, whereas the IMR–FDFD/FDFD took 48 s to converge to the full domain solution after three iterations, and for the same number of iterations, the IMR–FDFD/MoM took 40 s, on a 1.9-GHz P4 PC.

• • • •

CHAPTER 6

Parallelization of the Iterative Multiregion Technique

The purpose of this chapter is to present a simple introduction to the functionality and performance of a multiprocessor computer, which would help in understanding the idea of parallel programming. Once the idea is grasped, multiprocessors will be tied to the IMR technique to provide the advantage of speeding up the computational process, especially for problems of excessively large execution time, where the advantage of using number of processors will become more beneficial relative to a single processor. Further in this chapter, numerical analysis and results will be presented to show the performance of the IMR solution in conjunction with multiprocessors, showing an additional advantage of the IMR technique relative to a full domain problem simulation.

6.1   INTRODUCTION TO PARALLEL COMPUTING

The simplest definition to parallel computation is the situation where several processors simultaneously execute programs and cooperate to solve a given problem. Based on the resources provided to this work, all the experimental results will be performed on a single machine that contains several processors that communicate reliably and predictably. In this case, there is a single processor that distributes the tasks to every processor at the beginning of the computation and then gathers all the results and information at the end. In addition, each processor (P) has its own memory (M); it is called a distributed memory system and is illustrated in Figure 6.1. As for the execution process and the communication between processors, a synchronized model takes place between the processors. In a synchronous model, there usually are phases during which the processors carry out instructions independently from the others, where the communication between processors takes place at the end of a phase insuring that the next phase starts at each processor with updated information. That is the reason behind the overhead in time since each processor has to wait for the data to be received from the other processors. Despite the time overhead due to the communication process, the functionality of parallel processing is more efficient relative to a serial processor computation.

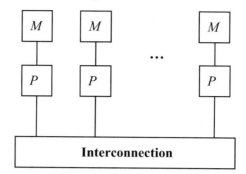

FIGURE 6.1: Distributed memory and CPU system.

6.2   COMMUNICATION TASK

The communication process is one of the important tasks that one has to consider when dealing with multiple processors. When solving a problem using parallel programming, the total solution time becomes a summation of the spent time for computing and the time spent for communicating. Hence, to achieve the best performance from parallel computing, it is desired to keep the computation time to communication time ratio as large as possible. Although the concept is quite obvious, it is still considered a challenging task especially when a large number of processors are involved, which tend to reduce the computational time and increase the communication time. There are a number of communication tasks that a programmer can perform based on the given problem. These tasks can be summarized as follows:

1. single-node broadcast;
2. multinode broadcast;
3. single-node and multinode accumulation;
4. single-node scatter;
5. single-node gather;
6. total exchange.

All these communication tasks are linked through a hierarchy, and they are ranked based on their complexity. Of course, the complexity of the communication operations changes according to the connections between the processors. Further details can be found and studied in many resources [42, 43].

6.3   HYBRID IMR—PARALLEL PROCESSING TECHNIQUE

Over the last decade, high-performance computing has been achieved through multiprocessing. Many computations can thus benefit from faster execution on parallel processors. Recently, MIT

Lincoln Laboratory has developed a way for implementing parallel computation using Matlab, version 7 [44]. Instead of writing an entire new application of Matlab, multiple Matlab applications can run simultaneously to share the computational load of a single Matlab program, where the multiple applications can communicate through some shared files. The Matlab message passing interface (MatlabMPI) model is designed for easy use on PCs with multiple processors.

In this chapter, the advantage of using multiple processors to speed up the solution of the IMR algorithm is suggested. Depending on the number of scatterers in the computational domain, a certain number of processors will be used. Each processor is responsible for performing the solution of each scatterer, where they will all work in parallel. The processors will communicate with each other in a way to accumulate the fields, calculated by each processor, and thus generate the final solution for the complete problem. For the ease of describing the algorithm, consider a scattering problem that consists of three separated objects. For this problem, three processors will be assigned, where each processor will compute the solution for one of the scatterers. Following the procedure of the IMR technique, once each processor calculates the electric and magnetic current densities, the processors start to communicate with each other sending the current information to the other processors. Each processor will thus perform the near-field calculations and send these data to the other processors, where new currents are generated at each processor based on the new excitation. After a number of iterations, the final solution is generated with an expected time saving relative to the solution generated using one processor for solving three subregions in series. One can expect for the suggested problem that a one-third time saving could be achieved, but because of the overhead caused by the communications between the processors, as described in the previous sections, a speed up factor less than three should be expected in this test example. Despite the communication process between the processors, significant time saving can be achieved for larger problems that can be divided into more than two subregions when using more than two processors on a single or multiple computers.

6.4 NUMERICAL RESULTS

In this section, numerical results are presented to show an additional advantage behind using the IMR technique in providing the flexibility of using multiprocessors, thus speeding up the total computational time relative to the full domain simulation of the same problem. Figure 6.2 shows a 2D scattering problem that consists of a dielectric slab of relative permittivity equal to 2.2 and relative permeability equal to 1, with width equal to 3λ and length equal to 15λ. The dielectric slab is placed 1.5λ away from a circular conducting cylinder of radius equal to 2.5λ. The parameter λ is the free-space wavelength that is assumed to be equal to 1 m. This configuration is excited by a TM_z plane wave incident from the positive x axis ($\phi^i = 0°$). The structure is first solved as a full domain

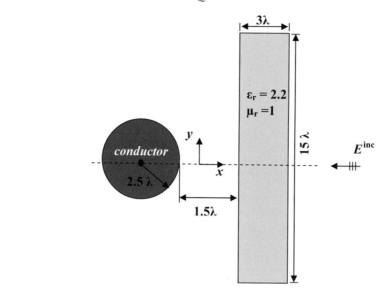

FIGURE 6.2: Problem description.

problem using the FDFD numerical technique, where the scattered near-field distribution (Figure 6.3) and the echo width (Figure 6.4) are computed. Figure 6.4 shows a comparison between the far field results generated from the full domain problem solution and that generated from the IMR technique after zero and four iterations. Good agreement between the full domain results and the IMR technique after four iterations is clearly observed.

Computational time comparison is performed to illustrate the advantage of using multiple processors in solving the problem defined in Figure 6.2. The full domain problem was first solved on a single processor using a 64-bit, 3.6-GHz Xeon processor with 8-GB RAM. The same problem was then solved using the IMR technique, on the same machine, using a single processor, and the total computational time was recorded after four iterations. As mentioned earlier, one of the advantages of using the IMR technique relative to a full domain simulation of a problem is that the first provides the ability of using multiple processors, which in turn reduces the total computational time. Thus, to illustrate this statement, the problem defined in Figure 6.2 was solved using the IMR technique where two processors were used; each was assigned to solve one of the subregions of the entire problem domain, and the total computational time was recorded after four iterations. Table 6.1 shows a computational time comparison between the three cases: 1) full domain simulation, 2) IMR solution on a single processor after four iterations, and 3) IMR solution on two processors after four iterations. Based on the data shown in Table 6.1, it can be clearly noticed that using the IMR

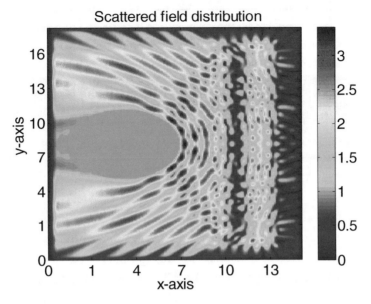

FIGURE 6.3: Scattered field distribution of the problem defined in Figure 6.2.

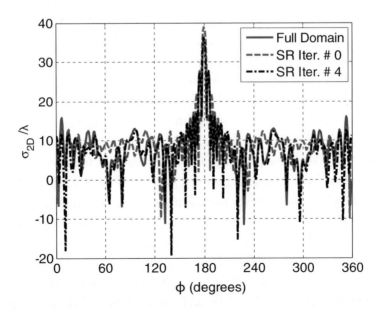

FIGURE 6.4: The echo width of the scattering geometry of Figure 6.2.

TABLE 6.1: Computational time comparison between full domain simulation, IMR solution on a single processor after four iterations and IMR solution on two processors after four iterations for the problem illustrated in Figure 6.2

	FULL DOMAIN	IMR: ONE PROCESSOR	IMR: TWO PROCESSORS
Computational time (min)	3.25	7.24	4.12

technique with two processors provided a 36% time saving relative to the IMR solution on a single processor. In addition to using multiple processors in conjunction with the IMR technique to speed up the total computational time of a problem, for this specific case, a memory saving of 43% was achieved using the IMR technique relative to the full domain simulation. This is because the IMR technique eliminates the extra memory used to simulate the space between the unconnected objects within the computational process. Furthermore, it also provides the flexibility of using different discretization in each of the subregions, where a discretization of 0.05λ was used in the subregion simulating the dielectric slab, whereas a different discretization of 0.075λ was used in the subregion simulating the conducting cylinder, in both x and y directions. This is relative to a discretization of 0.05λ used for the full domain simulation of the problem in both x and y directions.

Figure 6.5 shows a 3D scattering problem that consists of a dielectric rod of relative permittivity equal to 2.2 and relative permeability equal to 1, the dimensions of the rod are defined in Figure 6.5. The dielectric rod is placed 0.5λ away from a conducting ellipsoid of semiaxis; $Ra = 0.6\lambda$, $Rb = 0.4\lambda$ and $Rc = 0.4\lambda$, along the x, y, and z axes, respectively. This configuration is excited by a θ polarized, plane wave with $\phi^i = 90°$, and $\theta^i = 90°$. A cell size of 2 cm is used in the x, y, and z directions for both the full domain simulation and the IMR simulation of this problem, leading to a total number of cells of 1,302,048 for the full domain problem and a total number of cells of 833,856 for both regions using the IMR technique. Furthermore, to illustrate the advantage and applicability of using multiple processors to solve a 3D problem, precisely the one defined in Figure 6.5, the full domain configuration of this problem was first solved on a single processor using a 64-bit 3.6-GHz Xeon processor with 8-GB RAM. The same problem was then solved using the IMR technique, on the same machine, using a single processor, and the total computational time was recorded after three iterations. Because the goal of this section is to point out the advantage of using multiple processors in conjunction with the IMR technique, the same problem was again solved using the IMR technique using two processors working in parallel instead of one. Each was assigned to solve one of the subregions of the entire problem domain, and the total computational time was recorded after

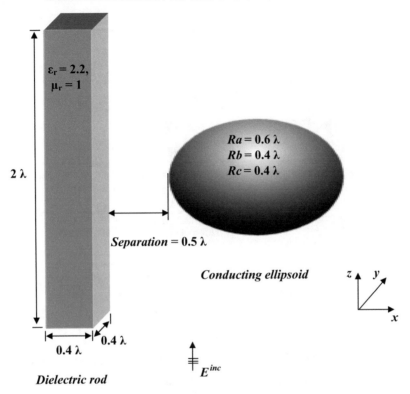

FIGURE 6.5: Geometry of a dielectric rod and a conducting sphere and the direction of the incident plane wave.

three iterations. The total computational time of the full domain problem was found to be equal to 58 min, whereas that of the IMR solution on a single processor, after three iterations, is equal to 62 min relative to 44 min recorded using the IMR solution computed on two processors.

Figure 6.6 shows the bistatic echo width for the problem defined in Figure 6.5 for the three plane cuts *xy*, *yz*, and *xz*, where a comparison between the far-field results generated using the full domain simulation of the problem and the far-field results computed using the IMR technique after zero and three iterations is presented. Good agreement between the full domain results and the IMR technique after three iterations is observed.

Table 6.2 shows a computational time comparison for the problem illustrated in Figure 6.5, similar to what was presented in Table 6.1, between the three cases: 1) full domain simulation, 2) IMR solution on a single processor after three iterations, and 3) IMR solution on two processors after three iterations. Based on the data shown in Table 6.2, it can be clearly noticed that using the IMR technique with two processors provided a 30% time saving relative to the IMR solution

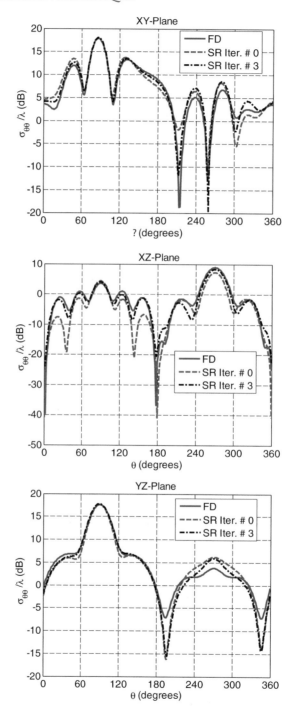

FIGURE 6.6: Bistatic echo width for *xy*, *xz*, and *yz* plane cuts.

TABLE 6.2: Computational time comparison between full domain simulation, IMR solution on a single processor after three iterations and IMR solution on two processors after three iterations for the problem illustrated in Figure 6.5

	FULL DOMAIN	IMR: ONE PROCESSOR	IMR: TWO PROCESSORS
Computational time (min)	58	62	44

on a single processor and a 24% time saving relative to the full domain simulation of the problem. In addition to using multiprocessors in conjunction with the IMR technique to speed up the total computational time of a problem, for this specific case, a memory saving of 36% was achieved using the IMR technique relative to the full domain simulation, where the same discretization was used in both solutions (i.e., the full domain solution and the IMR solution).

• • • •

CHAPTER 7

Combined Multigrid Technique and IMR Algorithm

The purpose of this chapter is to present a simple introduction to the MG technique pointing out the advantage of using this method in providing algorithms which can be used to accelerate the convergence rate of iterative methods, such as GMRES or BICGSTAB. This is done by providing these iterative methods with an approximate guess for the solution. This will enhance the solution process of the FDFD method, hence speeding up the computational process of the IMR algorithm. In addition to the MG technique, the ILU decomposition will also be used in this chapter as a preconditioning used to speed up the calculations. As a result of the use of both the MG technique and the ILU decomposition, in addition to the advantages of the IMR algorithm, it is possible to analyze large scattering problems with reasonable computer resources. In this chapter, the presented MG technique is used to solve 2D electromagnetic scattering problems, whereas the ILU preconditioner is described and used for both 2D and 3D problems.

7.1 INTRODUCTION TO MULTIGRID TECHNIQUE

When dealing with large-size problems, it is sometimes impossible to fit the number of unknowns in one matrix form and perform a direct matrix inversion solution. Thus, different iterative solvers were introduced to solve such problems, and they proved their efficient performance regarding memory and time. The MG technique is considered as one of those iterative solvers that can be used to solve this type of problems. It was widely spread starting from the early 1970s by Brandt [45], where it was introduced to perform a fast numerical simulation to the solution of boundary value problems. The MG method is an efficient technique generally used for solving smooth PDEs [46–52]. Initial interest in the MG method was geared toward overcoming the slow convergence rate of the classical iterative methods by updating blocks of grid points. Because of its superior performance, the MG technique is used in many applications such as solving the problem of huge power grids involved in VLSI designs that are required to distribute large amounts of current [53]. It is also utilized in the computation of gravitational forces together with a local refining mesh strategy [54]. In this chapter, the MG technique in conjunction with the IMR algorithm is developed and presented to solve large electromagnetic scattering problems, where the solution technique is based on the FDFD method.

The MG technique is based on dividing the computation domain into a number of levels starting from finer to coarser. The solution process is performed on the coarser level, then interpolation and correction steps take place until the solution at the finer grid level is achieved. This accelerates the rate of convergence of the iterative solvers used and thus speeds up the solution process. A remarkable CPU time saving can be achieved using the MG technique, as demonstrated by the presented numerical results, specifically for 2D problems. Further elaboration on the MG technique will be presented in the next sections.

7.2   MULTIGRID V-CYCLE ON A GRID HIERARCHY

The MG technique, as an iterative solver, is considered as a robust way to provide a solution to large-size problems that cannot be solved using a direct matrix inversion solution. The basic idea of the MG technique is to divide the computational domain into a number of levels (L), going from fine to coarser grid levels, as shown in Figure 7.1. The total number of grid points at each level (N_L) is taken to be $N_L = 2^L + 1$ for a square domain. At each level, a different discretization is used that is related to the discretization at the preceding finer grid level by $\Delta_{finer}/2$ for uniform meshing, where Δ_{finer} is the discretization used at the finer grid level. In each level, a relaxation scheme based on an iterative solver is introduced to smoothen out the errors. Gauss–Seidel and Jacobi methods are widely

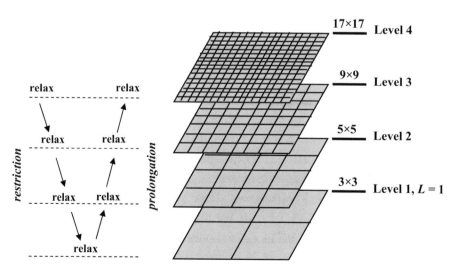

FIGURE 7.1: Multigrid V-cycle on a grid hierarchy.

used as relaxation schemes associated with the MG technique, where they proved their efficiency and high rate of convergence. The only drawback of these two iterative solvers is that they require a necessary condition for the solution to converge. That is, the diagonal elements of the coefficient matrix of the system of linear equations describing the problem have to be dominant over the off-diagonal ones. This is not the case for most of the numerical methods used to simulate large electromagnetic frequency domain problems based on differential type solutions. Thus, BICGSTAB is used instead as the relaxation scheme for the constructed MG technique presented in this work, where the solution generated at each level is considered as an initial guess to the BICGSTAB solver helping in accelerating the convergence rate.

Two basic operations are required in the MG technique to go through different levels: a restriction operator and a prolongation operator. The restriction operator is used to map the data onto a coarser gird level, whereas the prolongation operator maps the data from coarser to finer grid level based on a cubic analogue. The solution is finally performed on the coarser grid level at a very low computational cost due to the minimized domain size, where the number of required operations has an $O(N)$ computational cost, where N is the total number of grid points. Once the solution is computed at the coarsest level, the prolongation operation takes place in addition to some smoothing steps to interpolate the solution to the finest level. Finally, the desired solution is achieved in less computational time relative to other solution techniques.

7.3   MULTIGRID ALGORITHM

In Section B, Chapter 7, it was mentioned that the MG technique is used as an iterative solver to find an efficient solution for a set of linear equations that cannot be solved using a direct matrix inversion. This set of linear equations takes a matrix form of $[C][E] = [F]$, where $[C]$ is the coefficient matrix generated from the FDFD method; $[E]$ is the unknown scattered electromagnetic fields;

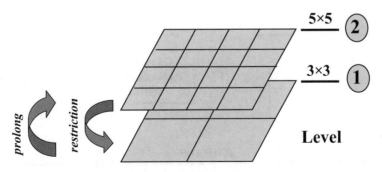

FIGURE 7.2: Example of a two-grid improvement scheme.

and $[F]$ is the excitation vector. Based on this matrix form generated for these linear equations, construction steps for the MG technique on a two-grid improvement scheme shown in Figure 7.2 are as follows:

1) Restrict $F_{(1 \times 25)}$ (level 2) to $F_{(1 \times 9)}$ (level 1).
2) Using a direct solution (direct inversion or Gauss elimination), find the solution for $E_{(1 \times 9)}$ (level 1) $= C_{(9 \times 9)}{}^{-1} F_{(1 \times 9)}$.
3) Prolong $E_{(1 \times 9)}$ (level 1) to get $E_{(1 \times 25)}$ (level 2).

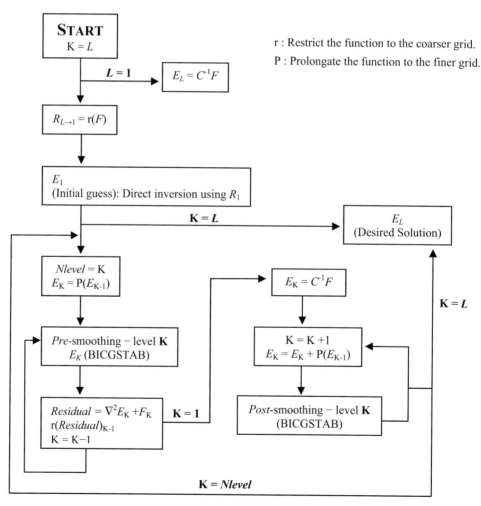

FIGURE 7.3: Block diagram of a single step within the multi-grid iteration. Reproduced/modified by permission of *ACES Newsletter*.

4) Pre-smooth (i.e., apply a relaxation scheme to smooth out the errors) $E_{(1 \times 25)}$ (level 2) using the BICGSTAB iterative solver. This is done by using the computed $E_{(1 \times 25)}$ from step 3 as an initial guess for the BICGSTAB function to provide a more enhanced values for $E_{(1 \times 25)}$ at level 2.

5) Calculate the residual $[C][E_{(\text{level } 2)}]$ $[F] = R_{(1 \times 25)}$.

6) Restrict the residual $R_{(1 \times 25)}$ to the coarser grid (level 1) $R_{(1 \times 9)}$.

7) Apply a direct solution for the solution correction $\Delta E_{(1 \times 9)} = C_{(9 \times 9)}^{-1} R_{(1 \times 9)}$.

8) Correction step: $E_{(1 \times 25)\text{new}} = E_{(1 \times 25)}$ (generated from step 3) + prolong $(\Delta E_{(1 \times 9)})$.

9) Post-smooth to $E_{(1 \times 25)}$ using a relaxation scheme based on the BICGSTAB iterative solver. Again this is done by using the $E_{(1 \times 25)\text{new}}$ computed at step 8 as an initial guess for the BICGSTAB function to provide a more enhanced values for $E_{(1 \times 25)}$ at level 2.

10) Required $E_{(1 \times 25)}$ at level 2 is achieved.

Figure 7.3 provides a block diagram describing the construction of the MG technique. In this book, unless otherwise specified, a two-level MG algorithm was used, with *V*-cycle as shown in Figure 7.2.

7.4   PRECONDITIONING

Finding a good preconditioner to solve a given sparse linear system is often considered a difficult but an important task. A preconditioner can be defined as any form of implicit or explicit modification, which leads to a new set of equations for a linear system which can then be easily solved using one of the iterative solvers. One of the simplest ways of defining a preconditioner is to perform an ILU decomposition of the original matrix C. This provides a decomposition of the form $C = LU$ R, where L and U refer to the lower and upper triangular matrices, respectively, and R is the residual of factorization. The ILU factorization is considered as an easy and inexpensive preconditioner to use. The more accurate the ILU factorization, the fewer the number of iterations will be required for the iterative solver to converge. The only drawback of the ILU preconditioner is the memory consumption that it requires to generate the two sparse matrices: L and U. Further information on this process can be found in [55].

In this book, the ILU preconditioner is used to solve both 2D and 3D scattering problems. It was claimed in [55] "*The incomplete LU factorization techniques were developed originally for **M**-matrices which arise from the discretization of partial differential equations of elliptic type, usually in one variable*" that the ILU preconditioner can function properly when being used to solve for one unknown, which would be the case for a 2D problem with one unknown scattered electric field component. But when the ILU preconditioner had been used to solve a 3D problem, it resulted in an ill-conditioned case. One can tie this with the statement of Yousef [55], since for a 3D problem,

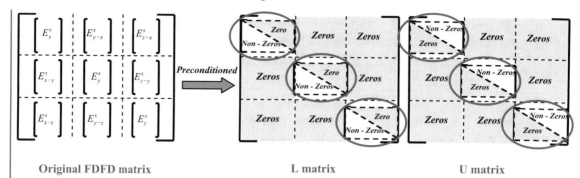

Original FDFD matrix   L matrix   U matrix

FIGURE 7.4: Efficient preconditioned matrix scheme for a matrix constructed from a 3D problem.

the unknown vector is a combination of the three scattered electric field components as shown in Chapter 3. Thus, for 3D problems, the used ILU factorization is a modified version of the traditional ILU to provide a stable and efficient solution.

The original matrix constructed from a 3D problem consists of nine submatrices; this is because we have three unknown field components to solve, each of size N, where N is the total number of nodes in the computational domain (i.e., $N = N_x \times N_y \times N_z$). Thus, the matrix size is of the order $3N$, so instead of generating a preconditioner for the original matrix, a preconditioned submatrix is created for the three diagonal submatrices as shown in Figure 7.4. This proposed reconfigured preconditioning scheme presented in this work does not just provide a stable and efficient preconditioner for 3D problems, but it also reduces the memory usage that a regular ILU matrix construction would require since all the off-diagonal submatrices are equal to zero.

7.5   NUMERICAL RESULTS

To validate the IMR–MG technique, a 2D problem is defined and sketched in Figure 7.5. Two rectangular dielectric cylinders of relative permittivity of 2.2 were embedded in an H-shaped conductor plate, and this structure is placed at 0.75λ away from a dielectric ellipse of relative permittivity equal to 3.4 with 0.435λ semiminor axis along the x axis and 1.1λ semimajor axis along the y axis. This configuration is excited by a TM_z plane wave incident from the negative y axis.

Shown in Figure 7.6 is a comparison between three solutions: the full domain results generated using the iterative solver BICGSTAB function, the FDFD solution of the full domain problem using the MG technique, and finally the IMR–MG algorithm computed after four iterations. The total number of cells used in the full domain problem is 63,504 cells, whereas that of the IMR solution for both regions is 18,285 cells, which is 71% less than that of the full domain. This is because different discretizations can be used in each of the subregions relative to the full domain

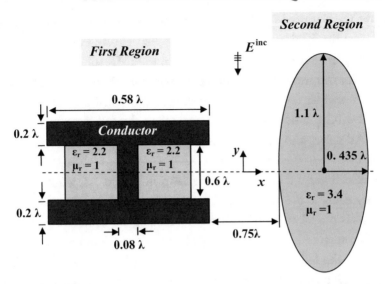

FIGURE 7.5: Problem configuration for two dielectric rectangular cylinders embedded in an H-shaped conductor next to a dielectric ellipse.

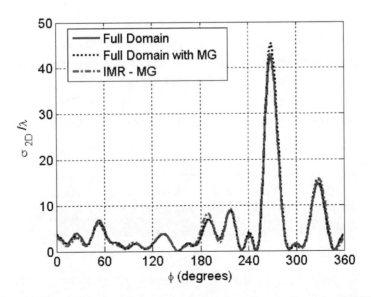

FIGURE 7.6: Comparison of the far field using the full domain solution using BICGSTAB, full domain solution solved using the MG technique, and the IMR–MG solution after four iterations.

problem, where the cell size used for the full domain simulation is 1 cm, whereas that of the IMR solution is 1 and 3 cm for the first and second regions, respectively. Furthermore, purging the cells used to simulate the gap between the scatterers within the IMR solution saves memory and time relative to the full domain simulation. Good agreement is achieved with significant reduction in memory size. A time analysis for the three approaches is shown in Table 7.1, where it can be clearly seen that the IMR–MG solution is 11 times faster than the full domain solution based on the iterative BICGSTAB solver and six times faster than the full domain solution based on the MG technique.

Another verification for a 2D problem is shown in Figure 7.7, where a dielectric ellipse of relative permittivity equal to 3.4 with 0.615λ semiminor axis along the x axis and 5λ semimajor axis along the y axis is placed at 0.75λ away from a rectangular conducting plate of length 0.58λ and width 1λ. This configuration is excited by a TM_z plane wave incident from the negative x axis. The cell size used for the full domain simulation is 2 cm, whereas that of the IMR solution is 2 cm for the first subregion and 2.175 cm for the second subregion in the x direction and 2.85 cm in the y direction. The structure is first solved as a full domain problem using the FDFD numerical technique, where the total near-field distribution (Figure 7.8) and the echo width (Figure 7.9) are computed. Figure 7.9 shows a comparison between the far-field results generated from the full domain problem solution and the IMR–MG technique after zero and four iterations. Good agreement can be clearly observed between the full domain results and the IMR–MG technique results after four iterations, with a 54% memory saving using the IMR technique relative to the full domain simulation of this problem.

Computational time comparison is performed to illustrate the advantage of using a parallel platform in solving this problem using the IMR technique with four iterations. The problem was first solved on a single processor using the BICGSTAB iterative solver having ILU factorization as a preconditioning to speed up the convergence rate of the solver. The problem was then solved using two processors where each was assigned to solve one of the subregions of the entire problem

TABLE 7.1: Time analysis for the three proposed approaches

SOLUTION	TIME (M)
Full domain: iterative solver	12
Full domain with multigrid	7
IMR–multigrid	2

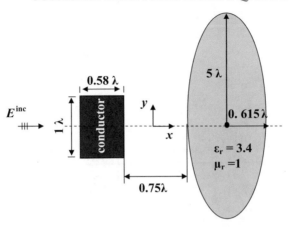

FIGURE 7.7: Problem configuration for a rectangular conductor plate placed next to a dielectric ellipse.

domain. Because the overhead caused by the communications between the processors, a speed up factor of 30% is achieved for this example.

To show the advantage of using the MG technique, the configuration depicted in Figure 7.7 is solved using the IMR algorithm on two processors, with and without the MG technique, to record the computational time difference. A time saving of 34% is achieved using the MG technique

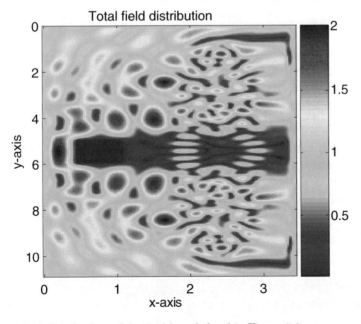

FIGURE 7.8: Total field distribution of the problem defined in Figure 7.7.

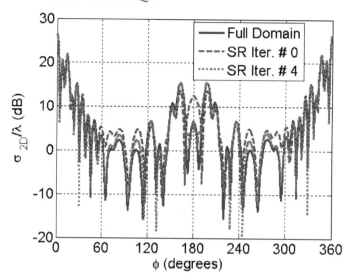

FIGURE 7.9: The echo width of the scattering geometry of Figure 7.7.

for this specific example. Even when the ILU preconditioning was used, which accelerates the rate of convergence of the BICGSTAB iterative solver, the MG technique proved its efficiency with a time saving of 2%. The time saving for this case, with the ILU preconditioning, is not significant because of the high rate of convergence of the BICGSTAB function in addition to the overhead in the time required by the MG technique. For the configuration depicted in Figure 7.7, a total

	SOLUTION	TIME (MIN)	SOLUTION	TIME (M MIN)	TIME SAVING (%)
1	IMR_Parallel (BICGSTAB_ILU)	5:42	Full domain (BICGSTAB_ILU)	8:25	30
2	IMR_MG (BICGSTAB)	10:26	Full domain (BICGSTAB)	15:55	34
3	IMR_MG (BICGSTAB_ILU)	8:02	Full domain (BICGSTAB_ILU)	8:25	2
4	IMR_Parallel_MG (BICGSTAB_ILU)	4	Full domain (BICGSTAB_ILU)	8:25	52

TABLE 7.2: Time analysis for the four proposed approaches for the configuration depicted in Figure 7.7

computational time saving of 52% is achieved using the parallel IMR–ILU–MG technique relative to that of a full domain solution. All simulations in this investigation were performed on a 64-bit 3.6-GHz Xeon processor with 2-GB RAM addressable by Matlab. Only 4 min 32 s is required to obtain the solution for this configuration using the parallel IMR–ILU–MG algorithm. This time comparison is clearly illustrated in Table 7.2.

•   •   •   •

CHAPTER 8

Concluding Remarks

In this work, an IMR technique is proposed to solve large-scale electromagnetic problems that can be decomposed into separate subregions using the FDFD method. This procedure starts by dividing the original computational domain into separate subregions where the solution is easily performed by the FDFD method followed by an iterative interaction process between the subregions. The new approach proposed here is found to be efficient in producing accurate results for the original problem with remarkable saving in both the total computational time and the computer memory usage especially if the separation between some subregions is large and/or coarser grids are used in some of the subregions, which may not be possible to use if only one domain is used for the solution of the problem. 2D and 3D problems are presented and tested to prove the concept of the IMR technique. Two techniques were proposed to speed up the calculations of the incident fields on the coupled domains: an averaging process and the use of the TF/SF technique, where the latter requires the computation of incident field components on a surface boundary rather than the entire computational domain. Therefore, the TF/SF formulation provided the IMR technique with the advantage of a remarkable time saving compared with the SF formulation.

A hybrid technique of two different solutions, MoM and FDFD, was proposed in this book to show the flexibility of the IMR technique in combining different solutions for each subregion to reach the desired solution in the most efficient way, by combining the desirable features of the two different numerical methods to analyze large-scale electromagnetic problems. The hybrid IMR approach proposed here is found to be efficient in producing accurate results for the provided test problems with more than 60% saving in the computer memory usage and with no significant change in the computational time. Furthermore, the idea of introducing multiprocessors to the IMR technique was proposed, which provided the advantage of speeding up the computational process, especially for problems of excessively large execution time, where the advantage of using a number of processors becomes more beneficial relative to a single processor. Without loosing generality, the advantage of using the IMR technique on a parallel platform is because the subregions are solved in parallel, each on a separate processor, leading to a remarkable CPU time saving. In addition to the idea of introducing multiprocessors to enhance the performance of the IMR technique, MG algorithm and ILU preconditioner were used in this work to accelerate the convergence rate of the

BICGSTAB iterative solver, which is essential especially when it is difficult or almost impossible to perform a direct matrix inversion solution to a large-size problem, resulting in speeding up the total computational time of the IMR technique.

Through this work, a number of enhancement procedures were conducted and combined with the main topic of this book, which is the IMR technique to provide a new robust and efficient approach to solve large-scale electromagnetic problems within reasonable time and with the least memory usage.

•   •   •   •

APPENDIX 1

Radiation and Scattering Equations

A.1 NEAR-FIELD 2D FORMULATION

The magnetic vector potential $A(x, y)$ produced by a TM_z line current source in a 2D space is given by

$$A_z(x, y) = \mu \iint_s J_z(x', y') \left[\frac{1}{4j} H_0^{(2)}(k_0 R) \right] ds' \qquad (A.1)$$

where the primed coordinates (x', y') represent the source, and the unprimed coordinates (x, y) represent the observation point. The intention here is to provide expressions for the electric and magnetic fields that are due to the magnetic vector potential of (A.1), which would be valid everywhere.

The magnetic field due to the potential of (A.1) is given as

$$H_A = \frac{1}{\mu} \nabla \times A = \frac{1}{4j} \nabla \times \iint_s J(x', y') \left[H_0^{(2)}(k_0 R) \right] ds'. \qquad (A.2)$$

Interchanging the integration and the differentiation as what was done in [29], we can write (A.2) as

$$H_A = \frac{1}{4j} \iint_s \nabla \times \left[J(x', y') \left(H_0^{(2)}(k_0 R) \right) \right] ds'. \qquad (A.3)$$

Using the vector identity

$$\nabla \times (gF) = (\nabla g) \times F + g(\nabla \times F), \qquad (A.4)$$

thus, (A.3) can be written as

$$\nabla \times \left[\left(H_0^{(2)}(k_0 R) \right) J(x', y') \right] = \nabla \left(H_0^{(2)}(k_0 R) \right) \times J(x', y') + H_0^{(2)}(k_0 R) \nabla \times J(x', y'). \qquad (A.5)$$

Because J is only a function of the primed coordinates and ∇ is a function of the unprimed coordinates, hence

$$\nabla \times J(x', y') = 0 \qquad (A.6)$$

and since

$$\nabla\left(H_0^{(2)}(k_0 R)\right) = -k_0 \hat{R} H_1^{(2)}(k_0 R),\tag{A.7}$$

where $\hat{R}$ is a unit vector directed along the line joining any point of the source and the observation point. Using (A.5) to (A.7) we can write (A.3) as

$$H_A(x, y) = \frac{-k_0}{4j} \iint_s \left(\hat{R} \times J\right) H_1^{(2)}(k_0 R)\, ds',\tag{A.8}$$

which can be expended in its two rectangular components in $x{-}y$ plane as

$$H_{Ax} = \frac{-k_0}{4j} \iint_s (y - y')\, J_z\, \frac{H_1^{(2)}(k_0 R)}{R}\, dx'\, dy',\tag{A.9}$$

$$H_{Ay} = \frac{-k_0}{4j} \iint_s (x - x')\, J_z\, \frac{H_1^{(2)}(k_0 R)}{R}\, dx'\, dy'.\tag{A.10}$$

Using Maxwell's equation, we can write the corresponding electric field components as

$$E_A = \frac{1}{j\omega\varepsilon}\nabla \times H_A\tag{A.11}$$

which can be reduced, using (A.9–A.10), to

$$E_{Az} = \frac{-k_0}{4\omega\varepsilon} \iint_s J_z \left[k_0 H_1^{(2)'}(k_0 R) + 2\frac{H_1^{(2)}(k_0 R)}{R} - \frac{H_1^{(2)}(k_0 R)}{R}\right] dx'\, dy'.\tag{A.12}$$

Similarly, the electric vector potential $F(x, y)$ based on M can be written as

$$F(x,y) = \varepsilon \iint_s M\left(x', y'\right)\left[\frac{1}{4j} H_0^{(2)}(k_0 R)\right] ds'\tag{A.13}$$

with the electric field component given by

$$E_F = -\frac{1}{\varepsilon}\nabla \times F = \frac{k_0}{4j} \iint_s \left(\hat{R} \times M\right) H_1^{(2)}(k_0 R)\, ds',\tag{A.14}$$

which can be expended in its rectangular component as

$$E_{Fz} = \frac{k_0}{4j} \iint_s \left[(x - x')\, M_y - (y - y')\, M_x\right] \frac{H_1^{(2)}(k_0 R)}{R}\, dx'\, dy'.\tag{A.15}$$

In the same manner, the corresponding magnetic field component can be written using Maxwell's equation

$$H_F = -\frac{1}{j\omega\mu}\nabla \times E_F \qquad (A.16)$$

which, with the help of (A.14), can be reduced to

$$H_{Fx} = \frac{k_0}{\omega\mu 4} \iint_s \left[\left[\frac{k_0 H_1^{(2)'}(k_0 R)(x-x')(y-y')}{R^2} - \frac{H_1^{(2)}(k_0 R)(x-x')(y-y')}{R^3}\right]M_y \atop + \left[\frac{-k_0 H_1^{(2)'}(k_0 R)(y-y')^2}{R^2} + \frac{H_1^{(2)}(k_0 R)(y-y')^2}{R^3} - \frac{H_1^{(2)}(k_0 R)}{R}\right]M_x \right] dx'\,dy', \qquad (A.17)$$

$$H_{Fy} = \frac{-k_0}{\omega\mu 4} \iint_s \left[\left[\frac{-k_0 H_1^{(2)'}(k_0 R)(x-x')(y-y')}{R^2} + \frac{H_1^{(2)}(k_0 R)(x-x')(y-y')}{R^3}\right]M_x \atop + \left[\frac{k_0 H_1^{(2)'}(k_0 R)(x-x')^2}{R^2} - \frac{H_1^{(2)}(k_0 R)(x-x')^2}{R^3} + \frac{H_1^{(2)}(k_0 R)}{R}\right]M_y \right] dx'\,dy'. \qquad (A.18)$$

A.2   FAR-FIELD 2D FORMULATION

The vector potentials in (A.1) and (A.13) can be rewritten for far-field expressions in the same manner as what was presented in [29] for 3D structures, as

$$A(x,y) = \mu \iint_s J(x',y')\left[\frac{1}{4j}H_0^{(2)}(k_0 R)\right]ds' = \frac{\mu H_0^{(2)}(k_0 r)}{4j}N\,, \qquad (A.19)$$

$$F(x,y) = \varepsilon \iint_s M(x',y')\left[\frac{1}{4j}H_0^{(2)}(k_0 R)\right]ds' = \frac{\varepsilon H_0^{(2)}(k_0 r)}{4j}L\,, \qquad (A.20)$$

where

$$N = \iint_s J_s\, e^{j\beta r'\cos\psi}\,ds'\,, \qquad L = \iint_s M_s\, e^{j\beta r'\cos\psi}\,ds'$$

and the large argument approximation for the Hankel function is given by

$$H_0^{(2)}(k_0 r)_{k_0 r\to\infty} = \sqrt{\frac{2j}{\pi k_0 r}}\, e^{-jk_0 r}.$$

The electric and magnetic field components can be written in the far field as

$$E_\theta = -\sqrt{\frac{jk_0}{8\pi r}}\, e^{-jk_0 r}\left[L_\phi + \eta N_\theta\right],$$

$$E_\phi = +\sqrt{\frac{jk_0}{8\pi r}}\, e^{-jk_0 r}\left[L_\theta - \eta N_\phi\right],$$

$$H_\theta = +\sqrt{\frac{jk_0}{8\pi r}}\, e^{-jk_0 r}\left[N_\phi - \frac{L_\theta}{\eta}\right],$$

$$H_\phi = -\sqrt{\frac{jk_0}{8\pi r}}\, e^{-jk_0 r}\left[N_\theta + \frac{L_\phi}{\eta}\right],$$

$$(A.21)$$

where N_θ, N_ϕ, L_θ, and L_ϕ are reduced to

$$N_\theta = \iint_s \left(J_x \cos\theta\cos\phi + J_y \cos\theta\sin\phi - J_z \sin\theta\right) e^{+j\beta r'\cos\psi}\, ds',$$

$$N_\phi = \iint_s \left(-J_x \sin\phi + J_y \cos\phi\right) e^{+j\beta r'\cos\psi}\, ds',$$

$$L_\theta = \iint_s \left(M_x \cos\theta\cos\phi + M_y \cos\theta\sin\phi - M_z \sin\theta\right) e^{+j\beta r'\cos\psi}\, ds',$$

$$L_\phi = \iint_s \left(-M_x \sin\phi + M_y \cos\phi\right) e^{+j\beta r'\cos\psi}\, ds'.$$

For a 2D TM_z problem, the only components involved in the calculation of the bistatic echo width are the E_θ and H_ϕ components, as the values of E_ϕ and H_θ are simply set to zero. Thus, E_θ here represents E_z from which the bistatic echo width can be calculated using the following formula

$$\sigma_{2-D} = \lim_{\rho\to\infty}\left[2\pi\rho\frac{|E_z^s|^2}{|E_z^i|^2}\right].$$

In the same manner, the corresponding magnetic field component can be written using Maxwell's equation

$$H_F = -\frac{1}{j\omega\mu}\nabla \times E_F \tag{A.16}$$

which, with the help of (A.14), can be reduced to

$$H_{Fx} = \frac{k_0}{\omega\mu 4}\iint_s \begin{bmatrix} \left[\dfrac{k_0 H_1^{(2)'}(k_0R)(x-x')(y-y')}{R^2} - \dfrac{H_1^{(2)}(k_0R)(x-x')(y-y')}{R^3}\right]M_y \\ + \left[\dfrac{-k_0 H_1^{(2)'}(k_0R)(y-y')^2}{R^2} + \dfrac{H_1^{(2)}(k_0R)(y-y')^2}{R^3} - \dfrac{H_1^{(2)}(k_0R)}{R}\right]M_x \end{bmatrix} dx'\,dy', \tag{A.17}$$

$$H_{Fy} = \frac{-k_0}{\omega\mu 4}\iint_s \begin{bmatrix} \left[\dfrac{-k_0 H_1^{(2)'}(k_0R)(x-x')(y-y')}{R^2} + \dfrac{H_1^{(2)}(k_0R)(x-x')(y-y')}{R^3}\right]M_x \\ + \left[\dfrac{k_0 H_1^{(2)'}(k_0R)(x-x')^2}{R^2} - \dfrac{H_1^{(2)}(k_0R)(x-x')^2}{R^3} + \dfrac{H_1^{(2)}(k_0R)}{R}\right]M_y \end{bmatrix} dx'\,dy'. \tag{A.18}$$

A.2   FAR-FIELD 2D FORMULATION

The vector potentials in (A.1) and (A.13) can be rewritten for far-field expressions in the same manner as what was presented in [29] for 3D structures, as

$$A(x,y) = \mu \iint_s J(x',y')\left[\frac{1}{4j}H_0^{(2)}(k_0R)\right]ds' = \frac{\mu H_0^{(2)}(k_0r)}{4j}N, \tag{A.19}$$

$$F(x,y) = \varepsilon \iint_s M(x',y')\left[\frac{1}{4j}H_0^{(2)}(k_0R)\right]ds' = \frac{\varepsilon H_0^{(2)}(k_0r)}{4j}L, \tag{A.20}$$

where

$$N = \iint_s J_s\, e^{j\beta r'\cos\psi}\,ds', \qquad L = \iint_s M_s\, e^{j\beta r'\cos\psi}\,ds'$$

and the large argument approximation for the Hankel function is given by

$$H_0^{(2)}(k_0r)_{k_0r\to\infty} = \sqrt{\frac{2j}{\pi k_0 r}}\,e^{-jk_0r}.$$

The electric and magnetic field components can be written in the far field as

$$E_\theta = -\sqrt{\frac{jk_0}{8\pi r}}\, e^{-jk_0 r}\left[L_\phi + \eta N_\theta\right],$$

$$E_\phi = +\sqrt{\frac{jk_0}{8\pi r}}\, e^{-jk_0 r}\left[L_\theta - \eta N_\phi\right],$$

$$H_\theta = +\sqrt{\frac{jk_0}{8\pi r}}\, e^{-jk_0 r}\left[N_\phi - \frac{L_\theta}{\eta}\right],$$

$$H_\phi = -\sqrt{\frac{jk_0}{8\pi r}}\, e^{-jk_0 r}\left[N_\theta + \frac{L_\phi}{\eta}\right],$$

$$(A.21)$$

where N_θ, N_ϕ, L_θ, and L_ϕ are reduced to

$$N_\theta = \iint\limits_{s} \left(J_x \cos\theta\cos\phi + J_y \cos\theta\sin\phi - J_z \sin\theta\right) e^{+j\beta r'\cos\psi}\, d s',$$

$$N_\phi = \iint\limits_{s} \left(-J_x \sin\phi + J_y \cos\phi\right) e^{+j\beta r'\cos\psi}\, d s',$$

$$L_\theta = \iint\limits_{s} \left(M_x \cos\theta\cos\phi + M_y \cos\theta\sin\phi - M_z \sin\theta\right) e^{+j\beta r'\cos\psi}\, d s',$$

$$L_\phi = \iint\limits_{s} \left(-M_x \sin\phi + M_y \cos\phi\right) e^{+j\beta r'\cos\psi}\, d s'.$$

For a 2D TM$_z$ problem, the only components involved in the calculation of the bistatic echo width are the E_θ and H_ϕ components, as the values of E_ϕ and H_θ are simply set to zero. Thus, E_θ here represents E_z from which the bistatic echo width can be calculated using the following formula

$$\sigma_{2-D} = \lim_{\rho\to\infty}\left[2\pi\rho\, \frac{|E_z^s|^2}{|E_z^i|^2}\right].$$

Bibliography

[1] B. Despres, "Domain decomposition method and the Helmholtz problem," in *Proceedings of International Symposium on Mathematical and Numerical Aspects of Wave Propagation Phenomena*. Strasbourg, France, 1992, pp. 44–52.

[2] B. Despres, "A domain decomposition method for the harmonic Maxwell equations," in *Iterative Methods in Linear Algebra*, R. Beauwens and P. de Groen, Eds. Amsterdam, The Netherlands: Elsevier, 1992, pp. 475–484.

[3] B. Stupfel and B. Despres, "A domain decomposition method for the solution of large electromagnetic scattering problems," *Journal of Electromagnetic Waves and Applications*, vol. 13, no. 11, pp. 1553–1568, 1999.

[4] B. Stupfel, "A fast-domain decomposition method for the solution of electromagnetic scattering by large objects," *IEEE Transactions on Antennas and Propagation*, vol. 44, no. 10, pp. 1375–1385, Oct. 1996. doi:10.1109/8.537332

[5] B. Stupfel and M. Mognot, "A domain decomposition method for the vector wave equation," *IEEE Transactions on Antennas and Propagation*, vol. 48, no. 5, pp. 653–660, May 2000. doi:10.1109/8.855483

[6] B. Stupfel, "A hybrid finite element and integral equation domain decomposition method for the solution of the 3-D scattering problem," *Journal of Computational Physics*, vol. 172, no. 2, pp. 451–471, Sept. 2001. doi:10.1006/jcph.2001.6814

[7] L. Yin and W. Hong, "A fast algorithm based on the domain decomposition method for scattering analysis of electrically large objects," *Radio Science*, vol. 37, no. 1, pp. 31–39, Jan.–Feb. 2002. doi:10.1029/1999RS002282

[8] L. Yin, J. Wang, and W. Hong, "A novel algorithm based on the domain-decomposition method for the full-wave analysis of 3-D electromagnetic problems," *IEEE Transactions on Microwave Theory and Techniques*, vol. 50, no. 8, pp. 2011–2017, Aug. 2002.

[9] P. Liu and Y.-Q. Jin, "The finite-element method with domain decomposition for electromagnetic bistatic scattering from the comprehensive model of a ship on and a target above a large-scale rough sea surface," *IEEE Geoscience and Remote Sensing*, vol. 42, no. 5, pp. 950–956, May 2004. doi:10.1109/TGRS.2004.825583

[10] J. Wang and W. Hong, "A fast-domain decomposition method for electromagnetic scattering analysis of 3-D objects," *2000 Asia-Pacific Microwave Conference*, pp. 424–427, 2000.

[11] Z. Qian, L. Yin, and W. Hong, "Application of domain decomposition and finite element method to electromagnetic compatible analysis," *IEEE Antennas and Propagation Society, AP-S International Symposium (Digest)*, vol. 4, pp. 642–645, 2001.

[12] W. Hong, X.X. Yin, X. An, Z.Q. Lv, and T.J. Cui, "A mixed algorithm of domain decomposition method and the measured equation of invariance for the electromagnetic problems," *IEEE Antennas and Propagation Society, AP-S International Symposium (Digest)*, vol. 3, pp. 2255–2258, 2004.

[13] L. Yin and W. Hong, "Domain decomposition method: a direct solution of Maxwell equations," *IEEE Antennas and Propagation Society, AP-S International Symposium (Digest)*, vol. 2, pp. 1290–1293, 1999.

[14] T. Horie, H. Kuramae, and T. Niho, "Parallel electromagnetic-mechanical coupled analysis using combined domain decomposition method," *IEEE Transactions on Magnetics*, vol. 33, no. 2, pp. 1792–1795, March 1997. doi:10.1109/20.582623

[15] C.T. Spring and A.C. Cangellaris, "Parallel implementation of domain decomposition methods for the electromagnetic analysis of guided wave systems," *Journal of Electromagnetic Waves and Applications*, vol. 9, no. 1–2, pp. 175–192, 1995.

[16] C.T. Wolfe, U. Navsariwala, and S.D. Gedney, "A parallel finite-element tearing and interconnecting algorithm for solution of the vector wave equation with PML absorbing medium," *IEEE Transactions on Antennas and Propagation*, vol. 48, no. 2, pp. 278–284, Feb. 2000. doi:10.1109/8.833077

[17] R. Lee and V. Chupongstimun, "A partitioning technique for the finite-element solution of electromagnetic scattering from electrically large dielectric cylinders", *IEEE Transactions on Antennas and Propagation*, vol. 42, pp. 737–741, May 1994. doi:10.1109/8.299575

[18] G.A. Thiele, "Overview of selected hybrid methods in radiating system analysis," *Proceedings of the IEEE*, vol. 80, no. 1, pp. 66–78, Jan. 1992. doi:10.1109/5.119567

[19] M. Carr and J.L. Volakis, "Domain decomposition by iterative field bouncing", *IEEE Antennas and Propagation Society, AP-S International Symposium (Digest)*, San Antonio, TX, vol. 3, pp. 298–301, 2001. doi:10.1109/APS.2002.1018214

[20] F. Xu and W. Hong, "Analysis of two dimensions sparse multicylinder scattering problem using DD-FDTD method," *IEEE Transactions on Antennas and Propagation*, vol. 52, no. 10, pp. 2612–2617, Oct. 2004. doi:10.1109/TAP.2004.834435

[21] A.Z. Elsherbeni, M. Hamid, and G. Tian, "Iterative scattering of a Gaussian beam by an array of circular conducting and dielectric cylinders", *Journal of Electromagnetic Waves and Applications*, vol. 7, pp. 1323–1342, 1993.

[22] Z. Wang, "A study of the numerical solution of two dimensional electromagnetic scattering problems via the finite difference method with a perfectly matched layer boundary condi-

tion," A Thesis in Master of Science in Engineering Science, Electrical Engineering Department, The University of Mississippi, 1995.

[23] A. Taflove and S.C. Hagness, *Computational Electrodynamics (The Finite-Difference Time-Domain Method)*. Boston: Artech House, Inc., 2000.

[24] A.Z. Elsherbeni, *Finite Difference Method Course Notes*, Electrical Engineering Department, The University of Mississippi, 2002.

[25] M.N.O. Sadiku, *Elements of Electromagnetics*. Orlando, FL: Holt, Rinehart and Winston, Inc., 1989.

[26] K.S. Yee, "Numerical solution of initial boundary value problems involving Maxwell's equations in isotropic media," *IEEE Transactions on Antennas and Propagation*, vol. 14, pp. 302–307, May 1966.

[27] G.L.G. Sleijpen and D.R. Fokkema, "Bicgstab(L) for linear equations involving unsymmetric matrices with complex spectrum," *Electronic Transactions on Numerical Analysis*, vol. 1, pp. 11–32, Sept. 1993.

[28] J.P. Berenger, "A perfectly matched layer for the absorption of electromagnetic waves," *Journal of Computational Physics*, vol. 114, no. 2, pp. 185–200, Oct. 1994.

[29] C.A. Balanis, *Antenna Theory (Analysis and Design)*. Arizona State University, AZ: John Wiley & Sons, Inc., 1982. doi:10.1109/5.119564

[30] A. Chatterjee, J.M. Jin, and J.L. Volakis, "Edge-based finite elements and vector ABCs applied to 3-D scattering," *IEEE Transactions on Antennas and Propagation*, vol. 41, no. 2, pp. 221–226, Feb. 1993. doi:10.1109/8.214614

[31] K. Umashankar and A. Taflove, *Computational Electromagntics*. Boston: Artech House, Inc., 1993, pp. 276.

[32] A.Z. Elsherbeni, M.H. Al Sharkawy, and S.F. Mahmoud, "Electromagnetic scattering from a 2D chiral strip simulated by circular cylinders for uniform and non-uniform chirality distribution," *IEEE Transactions on Antennas and Propagation*, vol. 52, no. 9, pp. 2244–2252. Sept. 2004.

[33] A.Z. Elsherbeni, "A comparative study of two-dimensional multiple scattering techniques," *Radio Science*, vol. 29, no. 4, pp. 1023–1033, July–August 1994. doi:10.1029/94RS00327

[34] B.J. Strait, *Applications of The Method of Moments to Electromagnetic Fields*. Syracuse, NY: The SCEEE PRESS, 1980.

[35] B. Stupfel, "A hybrid finite element and integral equation domain decomposition method for the solution of the 3-D scattering problem," *Journal of Computational Physics*, vol. 172, no. 2, pp. 451–471, Sept. 2001. doi:10.1006/jcph.2001.6814

[36] G.A. Thiele, "Overview of selected hybrid methods in radiating system analysis," *Proceedings of the IEEE*, vol. 80, no. 1, pp. 66–78, Jan. 1992. doi:10.1109/5.119567

[37] A. Monorchio, A.R. Bretones, R. Mittra, G. Manara, and R.G. Martin, "A hybrid time-domain technique that combines the finite element, finite difference and method of moment techniques to solve complex electromagnetic problems," *IEEE Transactions on Antennas and Propagation*, vol. 52, no. 10, pp. 2666–2673, Oct. 2004.

[38] J. Gong, J.L. Volakis, A. Woo, and H. Wang, "A hybrid finite element-boundary integral method for the analysis of cavity-backed antennas of arbitrary shape," *IEEE Transactions on Antennas and Propagation*, vol. 42, pp. 1233–1242, Sept. 1994.

[39] D. Jankovic, M. Labelle, D.C. Chang, J.M. Dunn, and R. Booton, "A hybrid method for the solution of scattering from inhomogeneous dielectric cylinders of arbitrary shape," *IEEE Transactions on Antennas and Propagation*, vol. 42, pp. 1215–1222, Sept. 1994.

[40] M.A. Mangoud, R.A. Abd-Alhameed, and P.S. Excell, "Simulation of human interaction with mobile telephones using hybrid techniques over coupled domains," *IEEE Transactions on Microwave Theory and Techniques*, vol. 48, pp. 2014–2021, Nov. 2000.

[41] M. Norgren, "A hybrid FDFD-BIE approach to two-dimensional scattering from an inhomogeneous biisotropic cylinder," *Journal of Electromagnetic Waves and Applications* (JEMWA), PIER 38, pp. 1–27, 2002.

[42] S. Brawer, *Introduction to Parallel Programming*. Boston: Academic Press, Inc., 1989.

[43] G. Pauletto, *Computational Solution of Large-Scale Macroeconometric Models*. Boston: Kluwer Academic Publishers, 1997.

[44] *Lincoln Laboratory*, website: http://www.ll.mit.edu/pMatlab/.

[45] A. Brandt, "Multi-level adaptive technique (MLAT) for fast numerical solution to boundary value problems," *Proceedings of the Third International Conference on Numerical Methods in Fluid Mechanics*, Paris, 1972.

[46] W.L. Briggs, *A Multigrid Tutorial*. Philadelphia, PA: SIAM, 1987.

[47] W. Hackbush, *Multi-Grid Methods and Applications*. Berlin, NY: Springer-Verlag, 1985.

[48] R.L. Haupt and S.E. Haupt, "An introduction to multigrid using Matlab," *Computer Applications in Engineering Journal*, vol. 2, no. 1. Jan. 1994.

[49] S.R. Nassif and J.N. Kozhaya, "Fast power grid simulation," *Design Automation Conference*, pp. 156–161. June 2000.

[50] W. Hackbusch, *Iterative Solution of Large Sparse Systems of Equations*. Berlin, NY: Springer-Verlag, 1995.

[51] J.H. Bramble, *Multigrid Methods*. Harlow, Essex, England: Longman Scientific & Technical, 1993.

[52] P. Wesseling, *An Introduction to Multigrid Methods*. Chichester, UK: John Wiley & Sons, 1992.

[53]   K.C. Sahu, *Numerical Computation of Spatially Developing Flows by Full-Multigrid Technique*, Masters Thesis, July 2003.

[54]   C. Gheller, F. Sartoretto, and M. Guidolin, "GAMMS: a multigrid–AMR code for computing gravitational fields," *ASP Conference*, vol. 314, pp. 670–673. 2004.

[55]   Y. Saad, *Iterative Methods for Sparse Linear Systems*. Philadelphia, PA: SIAM, 1996.

Author Biography

Mohamed H. Al Sharkawy was born in Alexandria, Egypt, in 1978. He graduated from the Department of Electrical Engineering, Arab Academy for Science and Technology (AAST), Alexandria, Egypt, on June 2000 and received both his Master of Science and Doctor of Philosophy degrees in electrical engineering from the University of Mississippi on October 2003 and December 2006, respectively. He worked as a postdoctoral research fellow in the Electrical Engineering Department at The University of Mississippi from December 2006 until June 2007. He was nominated a membership in the Sigma Xi society in 2004. He received The University of Mississippi Graduate Achievement Award in Electrical Engineering for 2005. He also received the best student paper award presented at the Applied Computational Electromagnetic Society (ACES) Symposium on March 2006. Dr. Al Sharkawy is a member of the Institute of Electrical and Electronics Engineers (IEEE) and the ACES. He has been serving as an assistant to the editor-in-chief for the *ACES Journal* since 2004. He has authored and coauthored more than 30 technical journals and conference papers. Dr. Al Sharkawy is currently an assistant professor at the Electronics and Communications Department at the Arab Academy for Science and Technology, Egypt. His research interests include electromagnetic scattering from parallel chiral and metamaterial cylinders and the application of finite difference time and frequency domain techniques for the analysis and design of antennas and microwave devices.

Veysel Demir was born in Batman, Turkey, in 1974. He received his Bachelor of Science degree in electrical engineering from the Middle East Technical University, Ankara, Turkey, in 1997. He received scholarship award from the Renaissance Scholarship Program for the graduate study in USA (2000–2004). He studied at Syracuse University, Syracuse, NY, where he received Master of Science degree in electrical engineering and Doctor of Philosophy degrees in 2002 and 2004, respectively. During his graduate study, he had worked as a research assistant in the Sonnet Software Inc., Liverpool, NY. He had been working as a visiting research scholar at the Electrical Engineering Department of The University of Mississippi from 2004 to 2007. He joined the Department of Electrical Engineering in Northern Illinois University in 2007, where he is working as an assistant professor. His research interests include numerical analysis techniques (FDTD, FDFD, and MoM) and microwaves and RF circuits analysis and design. Dr. Demir is a member of the IEEE and the ACES. He has authored and coauthored more than 20 technical journal and conference papers. He has been serving as a reviewer for the *ACES Journal* and *Transactions on Microwave Theory and Techniques*.

Atef Z. Elsherbeni received an honor Bachelor of Science degree in electronics and communications, an honor Bachelor of Science degree in applied physics, and a Master of Engineering degree in electrical engineering, all from Cairo University, Cairo, Egypt, in 1976, 1979, and 1982, respectively, and a Ph.D. degree in Electrical Engineering from Manitoba University, Winnipeg, Manitoba, Canada, in 1987. He was a part-time software and system design engineer from March 1980 to December 1982 at the Automated Data System Center, Cairo, Egypt. From January to August 1987, he was a postdoctoral fellow at Manitoba University. He joined the faculty at the University of Mississippi in August 1987 as an assistant professor of electrical engineering. He advanced to the rank of associate professor on July 1991 and to the rank of professor on July 1997. On August 2002, he became the director of the School of Engineering CAD Lab and the associate director of the Center for Applied Electromagnetic Systems Research (CAESR) at The University of Mississippi. He was appointed as adjunct professor at the Department of Electrical Engineering and Computer Science of the L.C. Smith College of Engineering and Computer Science at Syracuse University on January 2004. He spent a sabbatical term in 1996 at the Electrical Engineering Department, University of California at Los Angeles (UCLA) and was a visiting professor at Magdeburg University during the summer of 2005. Dr. Elsherbeni received the 2006 School of Engineering Senior Faculty Research Award for Outstanding Performance in research, the 2005 School of Engineering Faculty Service Award for Outstanding Performance in Service, the 2004 Valued Service Award from the ACES for Outstanding Service as 2003 ACES Symposium Chair, the Mississippi Academy of Science 2003 Outstanding Contribution to Science Award, the 2002 IEEE Region 3 Outstanding Engineering Educator Award, the 2002 School of Engineering Outstanding Engineering Faculty Member of the Year Award, the 2001 ACES Exemplary Service Award for leadership and contributions as electronic publishing managing editor 1999–2001, the 2001 Researcher/Scholar of the year award in the Department of Electrical Engineering, The University of Mississippi, and the 1996 Outstanding Engineering Educator of the IEEE Memphis Section. Dr. Elsherbeni has conducted researches dealing with scattering and diffraction by dielectric and metal objects, FDTD analysis of passive and active microwave devices including planar transmission lines, field visualization, and software development for EM education, interactions of electromagnetic waves with human body, sensors development for monitoring soil moisture, airports noise levels, air quality including haze and humidity, reflector and printed antennas and antenna arrays for radars, UAV, and personal communication systems, antennas for wideband applications, and antenna and material properties measurements. He has coauthored 90 technical journal articles and 21 book chapters, contributed to 240 professional presentations, and red 16 short courses and 18 invited seminars. He is the coauthor of the book entitled *Antenna Design and Visualization Using Matlab* (Scitech, 2006), the book entitled *MATLAB Simulations for Radar Systems Design* (CRC Press, 2003), and the main author of the chapters "Handheld Antennas" and "The Finite Difference Time Domain Technique for Microstrip Antennas" in the *Handbook of Antennas in Wireless Communications* (CRC Press, 2001). Dr. Elsherbeni is a fellow member of the IEEE. He is the editor-in-chief for *ACES Journal* and an associate editor to the *Radio Science Journal*. He serves on the editorial board of the *Book Series on Progress in Electromagnetic Research*, the *Electromagnetic Waves and Applications Journal*, and the *Computer Applications in Engineering Education Journal*. He was the chair of the Engineering and Physics Division of the Mississippi Academy of Science and was the chair of the Educational Activity Committee for the IEEE Region 3 Section. Dr. Elsherbeni's home page can be found at http://www.ee.olemiss.edu/atef, and his email address is elsherbeni@ieee.org.

Printed in the United States
by Baker & Taylor Publisher Services